Rodrigo Baldow

Brecht's Didactic Play and the Theatrical Text

Rodrigo Baldow

Brecht's Didactic Play and the Theatrical Text

Matuta and The Galileo Affair

ScienciaScripts

Imprint
Any brand names and product names mentioned in this book are subject to trademark, brand or patent protection and are trademarks or registered trademarks of their respective holders. The use of brand names, product names, common names, trade names, product descriptions etc. even without a particular marking in this work is in no way to be construed to mean that such names may be regarded as unrestricted in respect of trademark and brand protection legislation and could thus be used by anyone.

Cover image: www.ingimage.com

This book is a translation from the original published under ISBN 978-620-2-03232-2.

Publisher:
Sciencia Scripts
is a trademark of
Dodo Books Indian Ocean Ltd. and OmniScriptum S.R.L publishing group

120 High Road, East Finchley, London, N2 9ED, United Kingdom
Str. Armeneasca 28/1, office 1, Chisinau MD-2012, Republic of Moldova, Europe
Managing Directors: Ieva Konstantinova, Victoria Ursu
info@omniscriptum.com

Printed at: see last page
ISBN: 978-620-6-17243-7

SUMMARY

ACKNOWLEDGEMENTS

To my family for their support, affection, understanding and patience.

To **Professor Dr Jenner Barretto Bastos Filho** for sharing his knowledge, for his encouragement, for believing in this research and for his patience.

To **Professor Dr Alexandre José Gonçalves de Medeiros** for sharing his knowledge, for his encouragement, for believing in this research and for recommending and giving me the strength to do the Master's in Science and Mathematics Teaching at UFAL.

To my friend **Professor Edson Santos Júnior** for his great partnership.

To my friend **Professor Andréa de Souza e Silva** for her contributions to the construction of this work.

CHAPTER 1

INTRODUCTION

Science teaching in schools has not been very successful. Some pedagogical practices are one of the reasons why students haven't learned much, because they tend to be centred almost exclusively on setting exercises, which greatly limits their potential and even leads them to create a distorted image of the nature of science. As a result, many students lack interest in scientific knowledge (BATALHA et al, 2011).

When it comes to deepening and enriching scientific content, this should not only be understood in the sense of results and problem-solving; it is also necessary to look at the contexts of the "discoveries" of scientific theories, as well as their process of evolution. In this way, an attitude of this kind can provide the means to facilitate teaching that makes it possible to debate the workings and role of science, as well as seeing it as a historical entity (MOZENA, 2009).

Zanetic (2005) points out that contemporary citizens learn that physics has no relation to their daily lives and is not part of their culture. With the exception of a few isolated teachers who have experimented in the classroom, in the vast majority of cases physics is not taught well in schools and is restricted to teaching in which students memorise formulas and apply them mechanically to solve exercises that are typical of entrance exams. In order to improve the teaching of physics, it is necessary to work on its theoretical concepts, the history and philosophy of science, its relationship with society and how it relates to other cultures. In this way, there would be a contribution that would be more likely to provide a problematising, active and critical education.

The purely expository method, which has been used for a long time by most teachers, has not been very attractive to students. In view of this, there is a pressing need to look for new methodologies and pedagogical practices in Physics teaching that can awaken students' interest in this science. According to the recommendations put forward by the PCNs (BRASIL, 2002), different learning situations should be provided for students in Physics teaching so that they can fully develop their general skills and competences.

It is essential that schools work with different cultures, including scientific culture, thus developing in students a consistent view of science beyond simply conceiving of it as a mere set of contents and theories. In fact, scientific culture should be seen with its own rules, values and language, enabling students to understand, discuss and relate the content dealt with to physical phenomena and technological applications in a critical way (VIANNA, 2009).

When discussing culture, it's unlikely that anyone will talk about physical knowledge and its history. When the word culture is mentioned, people often think of a literary work, a symphony, a painting or something like that. Paintings by Picasso and Tarsila come to mind, as does a Beethoven symphony, a Bachiana by Villa Lobos, or a novel by Dostoevsky and Machado de Assis, and in popular culture capoeira, samba or a tango. But hardly anyone will use Maxwell's equations as an example of culture (ZANETIC, 2005).

Gardair and Schall talk about how to work with and view art, science and education:

> In the triad of art, science and education, art should not be seen as a mere resource, just as science should not be reduced to the content to be presented. More than that, like science, art is a constructor of points of view. Both are the fruit and expression of a given era and should be presented as such. Thus, the proposal to educate based on the interaction between science and art cannot dispense with the idea that these two ways of knowing and expressing the world are born out of man's need to seek answers to his inconclusiveness. Whether through poetic licence or scientific rigour, or even scientific licence and poetic rigour, man seeks answers to original questions. Who is he?
> Where does it come from? Where does it go? - these are some of the questions that have led man to build science, art and so many other ways of reading the world (GARDAIR; SCHALL, 2009, p. 710).

Science is not seen as part of culture, and physics is taught by the majority of teachers on the basis of fixing exercises, with students having the "task" of memorising formulas in order to apply them to solving questions with characteristics similar to those that appear in entrance exams. These circumstances contribute to what Cachapuz and collaborators (2005) call the mistaken image of scientific activity, which distorted scientific education itself has unfortunately reinforced through its action or omission. These authors identified seven deformed visions of science:

- **Decontextualised**: This is when science and technology are talked about without showing their relationship with society;
- **Individualist and Elitist**: This view shows science as the knowledge of a few. It is isolated geniuses who understand it. Just as scientific knowledge is discovered by isolated scientists;
- **Empirical-Inductivist and Atheoretical**: This is when they advocate learning science through experiments without theoretical knowledge. A neutral observation;
- **Rigid, algorithmic, infallible...**: And when they consider **the** "scientific method" as something exact that follows a sequence with experiments as essential for proof;
- **Aproblematic and ahistorical**: This is when science is taught as if it had "ready-made" knowledge, with absolute truths. Not showing the historical development of scientific knowledge. Not showing also that "all knowledge is the answer to a question" (BACHERLARD apud CACHAPUZ ET AL, 2005, p. 49) ;[1]
- **Exclusively Analytical**: This is the view of science based on reductionism in a partialised and simplified way without synthesis and studies of increasing complexity;
- **Accumulative**: This is when scientific development is presented as a linear growth. The problems encountered and the "fights" between rival theories are not presented.

Given these views, we sought answers to the following research problem: Does **a theatre approach to** The Galileo Affair**, based on Brecht's Didactic Play, contribute to a change, albeit partial, in the view of science as absolute truth?**

In an attempt to break through some warped views and bring art and science

[1] The authors based this last idea on: Bachelard, G.. The Formation of the Scientific Spirit. Paris: Vrin, 1938.

closer together, or to intertwine them, a theatre text was written about the moment that has become known in the history of science as The Galileo Affair, which shows the Florentine thinker's conflict with the Catholic Church, specifically during the years 1632 and 1633. During much of the play's writing, issues related to physics and astronomy that Galileo debated in his time will also be discussed.

In order to answer the research problem, the following objectives were set:

- **General Objective**

  Investigate how science theatre can contribute to changing students' view of science as an absolute truth, even if only partially.

- **Specific objectives**

  - Construct a theatre text about The *Galileo Affair* as a tool to be used in the debate about some warped views of science;

  - To analyse the students' view of the attitude and attitudes of the philosopher Galileo in the conflict with the Church and vice versa;

  - Identify whether the scientific theatre text favours the understanding of a science within a historical context.

Faced with the difficulties encountered in teaching physics and the misconceptions of scientific activity that science education has reinforced by its action or omission, we justify this research by the intention of contributing to changing some deformed views of science based on a pedagogical practice that worked with a Theatrical Text constructed and that addresses *The Galileo Case* with Brecht's Didactic Play as its theoretical foundation.

As a Theatrical Text was written, dealing with a historical moment in Physics, section 2 will deal with Theatre and the History of Physics in Physics Teaching. Section 3 will give a brief history of the philosopher Galileo Galileo, with an emphasis on the historical moment known as *The Galileo Affair.* Section 4 discusses the structure of David Ball's Theatrical Text and Brecht's Didactic Play. Section 5 shows the Methodology. In section 6 we'll analyse the play. And finally, the final considerations.

CHAPTER 2

THEATRE AND HISTORY OF PHYSICS IN PHYSICS TEACHING

2.1 History of Physics in Physics Teaching

The history of science is an extremely important teaching tool in science education for several reasons:

> ...as a way of introducing science in humanities courses; as a means of facilitating and deepening the conceptual understanding of science itself; as an instrument to allow the discussion and teaching of the very nature and method of science; and other purposes (MARTINS, 2000, p. 47).

Prestes and Caldeira (2009) point out that Matthews (1994) summarises some arguments showing the importance of the History of Science in Science Teaching:

> History promotes a better understanding of scientific concepts and methods.
>
> Historical approaches connect the development of individual thought with the development of scientific ideas.
>
> The history of science is intrinsically valuable. Important episodes in the history of science and culture - the scientific revolution, Darwinism, the discovery of penicillin, etc. - should be familiar to every student.
>
> History is necessary to understand the nature of science.
>
> History neutralises the scientism and dogmatism that are often found in science teaching manuals and classes.
>
> History, by examining the lives and times of individual researchers, humanises scientific material, making it less abstract and more interesting to students.
>
> History favours connections to be made within scientific topics and disciplines, as well as with other academic disciplines; history exposes the integrative and interdependent nature of human acquisitions (MATTHEWS apud PRESTES; CALDEIRA, 2009, p. 3 and 4).

According to Bastos Filho (2012), the use of the

History of Science, as well as Philosophy of Science, in the teaching of this area, particularly in Physics, providing a richer learning of the subject

approached. The inclusion and interweaving of the History of Physics with the

Philosophy of Physics in the teaching of this subject contributes to the understanding of the content of this area.

According to Mozena (2009), the history and philosophy of science should be part of physics teaching, since they lead students to understand the role of science in the past, as well as in the present in society. In this way, physics can be understood as culture.

Working with the History and Philosophy of Science in the classroom provides teaching that can develop skills in students, allowing them to understand the evolutionary processes of building knowledge in this area. In this way, students will be able to understand that science has had an evolutionary process, without seeing it as isolated points with no connection to everyday life (BATALHA, 2011).

Despite the importance of the History of Science in Science Teaching, the majority of didactic texts that do not work in some way with this line, are contributing to some deformed visions of Science. Baldow and Monteiro Jr (2010) and Gatti, Nardi and Silva (2010) highlight these flaws:

> The history of science is important for carrying out an epistemological analysis of the development of scientific theories. Ideas are not born from a simple snap of the fingers as, unfortunately, many books end up revealing. On the other hand, a factual history, centred on isolated facts or summaries of scientists' biographies, contributes very little or almost nothing to the teaching of science (BALDOW; MONTEIRO JR, 2010, p. 17).

> Most didactic texts on the History of Science present scientific knowledge as a linear progression of events, producing a cumulative and immutable knowledge, the result of discoveries made by brilliant scientists" (GATTI; NARDI; SILVA, 2010, p. 17).

Perez et al. (2001) reinforce this idea by stating that science teaching in higher education is practically reduced to the presentation of previously prepared knowledge, without investigative activities, not showing the problems that gave rise to it, the evolution, the difficulties encountered, etc.

Contributing to the students' image of science as unique, algorithmic, well-defined and infallible.

2.1 Theatre in Physics Teaching

Currently, most students are not interested in the subject of Physics taught in schools. This has happened due to various factors that can be considered demotivational. One of the main ones is teaching physics without showing how it relates to everyday situations. Due to this lack of motivation among students in wanting to learn this science, many teachers end up looking for different methods of teaching physics, finding, for example, in the theatre activity an excellent pedagogical practice that has made students enthusiastic about learning scientific knowledge. Some authors (MOURA; TEIXEIRA, 2008; SILVEIRA; ATAÍDE; FREIRE, 2009; GARDAIR; SCHALL, 2009; MEDINA; BRAGA, 2010, BATALHA ET AL, 2011; MACHADO; MATOS, 2012; MOREIRA, 2013) have written works that defend the use of theatre as a pedagogical practice in science teaching.

With regard to theatre, Brecht points out that:

> People go to the theatre to be seduced, enchanted, impressed, exalted, amazed, moved, captivated, freed, distracted, relieved, moulded, pulled out of their time, provided with illusions. All this is so natural that the definition of art is given precisely by the fact that it liberates, seduces, etc. It is no longer art if it doesn't do this (BRECHT, 1967, p. 136 and 137).

Gwendola states that:

> Man needs to understand the meaning of his existence on earth, to laugh and to cry. Theatre doesn't provide answers, but it is a means of concretely demonstrating that we are aware of life. [...] Thanks to art, we ask ourselves questions and become our own authors [...] Theatre puts the world on stage to help us understand it (GWENDOLA apud GARDAIR; SCHALL, 2009, p. 696).

Based on Barbosa (2002), Machado and Matos (2012) emphasise that the aim of combining art and education is to understand how the former is constituted as a specific manifestation of human creativity in interaction with the world in which they live. Theatre is a method that fits in very well with science teaching.

According to Medina and Braga (2010), theatre and theory have the same etymological origins . Both have the same Greek radical *thea*, which means "a point of view". In this way, it can be said that science has a theatricality all of its own, because the exercise of scientific activity can involve controversies, "fights", ambitions, arguments and their counter-arguments. These are good elements for a theatre play.

Medina and Braga (2010) also point out that theatre allows students to find themselves in a project, feeling understood and recognised. This activity also teaches them to live and work collectively, respecting others and their commitments, complying with rules and having discipline (with the timetable and fidelity to the text, for example). In this way, it helps students realise that individual success reflects on the success of the project. And this requires discipline and teamwork.

In the same article, Medina and Braga (2010), evaluating the work they did with students using theatre, addressing the history of science, stated that the relationship between the history and philosophy of science and theatre made them think about the different disturbing problems related to the human essence, such as ethics, honour, character, value, the use of scientific instruments, contradictions and obstacles to scientific development. This activity contributed to the relationship between everyday science and politics, emphasising paradigm shifts. Thus, the authors justified the use of theatrical resources as an activity capable of discussing scientific issues.

Silveira, Ataíde and Freire (2009) point out that from the development of two plays that they analysed, they realised that when science is communicated through theatre, it provides an activity that involves the students more, not only in the performance of the play, but also in the discussions and problematisations that took place on purpose from the characters' speeches. In addition, the theatre provided a more accessible language. They also point out that this activity can be the starting point for student interest, dissemination and popularisation in a playful way of teaching scientific knowledge.

Oliveira and Zanetic (2004) point out that in a theatre activity there is the possibility of working with students on sensitivity, perception and intuition. Addressing scientific content in this activity will enable students to relate the subject to social issues.

A theatre activity that deals with the history of physics provides students with a moment that can arouse their curiosity about the content covered, the scientists involved and the story itself, since this type of activity tends to be more creative and fun than a class that only works with mathematical equations (COSTA; RIBEIRO; SOUZA, 2005).

The relationship between theatre and science is not something new. At the same time, it is not "normal" to find plays that deal with scientific knowledge as their main subject. However, Gardair and Schall (2009) recall some good examples of theatre texts that make this interaction, such as Bertolt Brecht's "The Life of Galileo" (1956), Friedrich Dúrrenmatt's "The Physicists" (1960), Heinar Kipphardt's "The Oppenheimer Case" (1964), Araújo Porto Alegre's "The Amazon Statue" (1851), Machado de Assis's "Lesson in Botany" (1906); "Copenhagen" (1998), by Michael Frayn, and "The Test" (2000), a play written by David Auburn. Medina and Braga (2010) also point out that some great scientists have become theatre characters, such as Einstein (Einstein, 1998), Niels Bohr and Heisenberg (Copenhagen, 2001), Richard P. Feynman (And now Mr Feynman?, 2004), Newton and Leibniz (Calculus, 2003), Lavoisier (Oxygen, 2006) and Galileo (The Life of Galileo).

In relation to the theatre text "The Life of Galileo" written by Brecht, Vaz comments on it and also on the scientist:

> He (Galileo) participated in various spheres of the society in which he lived. Perhaps that's why Galileo receives so much attention to this day. And it's not just essays and treatises on the history and philosophy of science. References to him are recurrent in literature and in areas that are normally far removed from physics, such as political philosophy. In the play **Galileo Galilei** written by Bertolt Brecht, for example, he is portrayed as an intelligent, courageous but conflicted person. Four hundred years after he lived, Galileo strikes us as a great man, in that he was aware of the wisdom of those who worked in workshops and on building sites, while at the same time despising the haughtiness of the learned men of his time. Men of knowledge, but who refused to elaborate their own reasoning (VAZ, 2009, p. 180, emphasis added).

Luisa Massarani and Carla Almeida did an interview with actor Carlos Palmas, who spoke about the relationship between Art and Science that he made in his project entitled "Art and Science on Stage". Talking about how he began this relationship in the theatre, he recalled the day he was enchanted when he saw the monologue Einstein in Chile in 1995, written by the Canadian Gabriel Emanuel, and bought the copyright to the play and began showing it in São Paulo in 1998. About the text of this play, he said: "With the text in hand I went to see what Einstein meant by this and that, that's when I started reading and researching what I should have learnt at school" (PALMAS, 2006, p. 235). That's when Carlos Palmas created the "Art and Science on Stage" project. They have also staged the plays *Copenhagen*, by Englishman Michael Frayn, Perdida, uma comédia quântica, by Spaniard José Sanchis Sinisterra, Quebrando Códigos, by Englishman Hugh Whitemore, and E agora Mr Feynman?, by American Peter Parnell. In this interview, he spoke about the importance of approaching scientific knowledge in the theatre:

> Through artistic work, we can very well focus on scientific knowledge. I think it's plausible to bring more in-depth knowledge of reality, of man, of nature and of life into art. Science is part of man's culture; it goes hand in hand with everything that is done in art. There is a whole technology that collaborates with artistic production, whether in theatre, cinema or literature. I don't see any difficulty in bringing science to the centre of art (PALMA, 2006, p. 234).

Theatre has been gaining prominence in science museums in Brazil. Moreira (2013) states that this prominence is associated with the aims and objectives of these places, which need new ways of interacting with the visiting public in order to arouse the viewer's interest and attention. Theatre is a mass communication strategy that can inform, entertain and sensitise the visiting public to issues related to science. To this end, it is important that the play's plot presents questions to be resolved, addressing conceptual, philosophical and historical aspects of science and technology, especially in terms of conflicts that can be related to the spectator's everyday life in the play.

CHAPTER 3

A BRIEF HISTORY OF GALILEO GALILEI WITH AN EMPHASIS ON THE HISTORICAL MOMENT KNOWN AS THE GALILEO AFFAIR

Aristotle, a Greek philosopher who lived in the 4th century BC (384/383 BC to 322 BC), said that all things that existed in the region between the Earth and the Moon were made up of a combination of four elements: earth, water, air and fire. There was an organisation between these elements in the world, so that earth was at the bottom, water above it, then air and fire at the top. Because of this order, Aristotle was able to explain some "natural" movements. For example, when a stone was dropped, it would fall towards the centre of the Earth. This happened because the stone was made up mainly of the element earth, so it sought its natural place, which was below. Another example was when you lit a fire, the smoke would rise because it was looking for its natural place at the top, where the fire was. These natural movements were part of Aristotelian physics in the sublunary world (ZYLBERSTAJN, 2011). These natural movements of stone and smoke were downwards in the first case and upwards in the second, following a straight line that passed through the centre of the Earth. But there were also movements in other ways. For example, a stone could be thrown upwards. In this case, according to Aristotle, it would be making a *violent* movement. And this was only possible when a "force"[2] was acting on the body so that it maintained a movement contrary to what was natural (COHEN, 1967).

Cohen (1967) also points out that in the imperfect world of the Earth, Aristotelian physics said that when an object was made up predominantly of the element earth and/or water, it was heavier, and when it was made up of the element fire and/or air, it was lighter.

---

[2] It's important to emphasise that force for Aristotle means changing the order of the cosmos and is therefore necessarily something ephemeral because the cosmological order can only be violated for a short time. The ascent of a stone by the action of someone throwing it upwards is a violent movement only during the ascent, becoming natural again when the stone begins to fall towards its natural place, which is the centre of the Earth. There is no force there. It is clear that Aristotle's concept of force differs radically from the Newtonian concept of force (for more details see KOYRÉ, 2009).

Above we showed how Aristotelian physics explained the sublunar world. But in the sky, above the Moon, including it, the cosmology he defended presupposed the existence of the ethereal substance, qualitatively different from the other four substances of the sublunary world. In this supralunar world, everything was made of a fifth element called ether; all the bodies that were part of this ethereal "world" had a perfectly spherical and smooth shape. According to Cohen (1967), Aristotelian cosmology stated that the bodies of the supralunar world naturally moved in a circular motion around the Earth. And the element ether was immutable and incorruptible, creating the idea of a "perfect" sky. Since the Earth wasn't made of this element, it was "natural" that it didn't have the circular motion of orbiting another star or spinning around its own axis.

For Aristotle, the weight of the object influenced the speed of descent. For example, if two stones of different weights were dropped from the same height at the same moment, the heavier one would fall first. Following more modern thinking, it can be said that in Aristotelian physics the weight of the body was proportional to the speed[3] of its fall. And its natural movement was in a straight line.

But when Aristotle thought about it, he didn't do experiments to be sure in practice, since for him, according to Drake (1981), practical knowledge, which he called techne, was not considered scientific knowledge, calling the latter episteme. Scientific knowledge, episteme, was reason, with the aim of understanding things by means of their causes, and this is what was relevant to Aristotle in relation to science.

This idea that a body with more weight falls faster was one of the first doubts of the Italian Galileo Galilei (1564-1642) in relation to Aristotle's knowledge of the philosophy of nature, since he first saw stones of different sizes falling from the same height at the same moment and reaching the ground at almost the same time (DRAKE, 1981).

---

[3] Koyré problematises the mathematical use of Aristotelian physics, because it falsifies the spirit. In one of his footnotes, he addresses this problem: "Aristotelian physics is essentially non-mathematical, and it is not possible to mathematise it (presenting it, for example, as founded on the principle that speed is directly proportional to force and inversely proportional to resistance, a proportionality that is nothing more than the continuation of Aristotelian principles) without falsifying its spirit." (KOYRÉ, 2009, p. 8).

In 1604, Galileo made several measurements in which he let a ball roll down an inclined plane from rest. The inclination was less than 2°, and he used musical beats lasting about half a second to mark the distances travelled by the ball in equal times. In this experiment, Galileo realised that in the series of equal times, the simplified distances were increasing following a sequence of odd numbers 1, 3, 5, 7,... and when the distances were accumulated, they followed the sequence 1, 4, 9, 16. Observing these measurements, he arrived at the law of falling bass, which stated, in simplified form, that the distance travelled by the object was proportional to the square of the time of its fall (DRAKE, 1981).

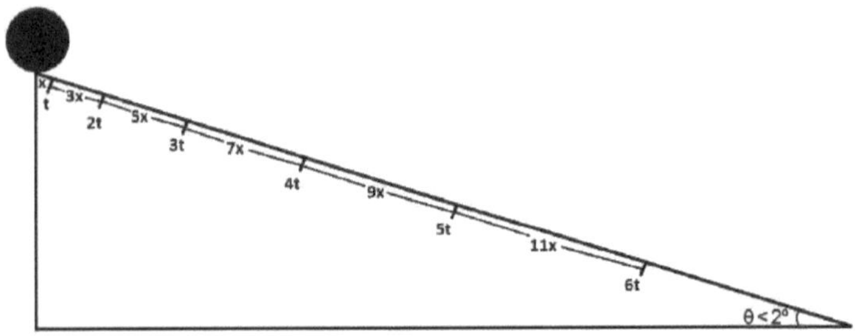

Fig. 01 - Inclined plane experiment and the relationship between the distance travelled and the square of the time.

Source: prepared by the author.

When Galileo began to use calculations in physics, he ended up attaching importance to this area, so that many believed that philosophy was being sidelined. By inventing a science that was not accepted in philosophy, he ended up creating friction with some people in this area (DRAKE, 1981). According to Santillana (1960), Galileo had great difficulty showing that mathematical studies were very important in the philosophical discussions of the time related to the sciences.

Galileo's ideas were not only different from those of Aristotle. Drake (1981) points out that even though Plato was a great advocate of maths, Galileo had a different view. Observe:

> Galileo's opinion on the place of maths in physics is as different from Plato's as it is from Aristotle's. Plato considered only the world of pure mathematical ideas to be worthy of study. Plato considered that only the world of pure mathematical ideas is worthy of study; if physical objects don't conform to it, so much the worse for them, because they are defective and imperfect in any way. Aristotle considered the procedures of maths to be foreign to physics, because mathematicians left matter completely out of the picture. Both philosophers were impressed by the abstract

> nature of maths, in contrast to the concrete material world. Galileo, on the other hand, was impressed by the usefulness of maths as a tool for studying physics. Precisely because calculus didn't fit exactly with observation, there was no reason to exalt one or abandon the other. A poor fit could show that we should leave maths or observations out of it (DRAKE, 1981, p. 87 and 88).

The geocentric theory of the Greek philosopher Claudius Ptolemy (1st and 2nd century AD), based on the ideas of Aristotle, admitted two movements of the stars. Firstly, they went round the Earth in a circular path, as shown in the circle in fig. 02, which has a point P. Around this point, the planets made another circular rotation. The latter movement was called the Epicycle and the former the Deferent.

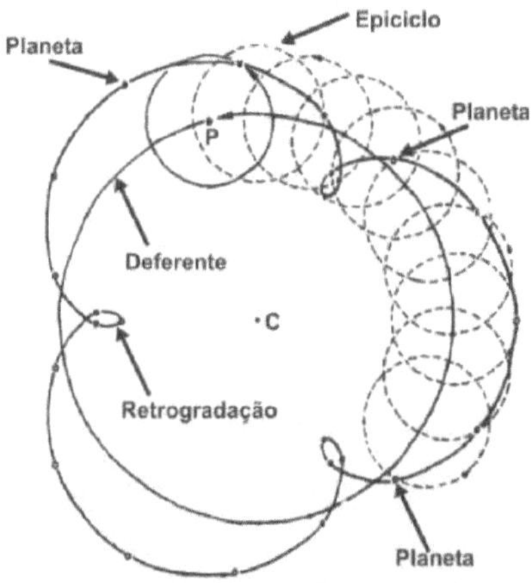

Fig. 02 - Movement of the Stars Around the Earth According to the Ptolemaic Theory

Source: http://www.ccvalg.pt/astronomia/historia/antiguidade/epicilo_deferente.gif

Ptolemy didn't believe in the Earth's rotation, because if it did, Aristotelian physics would be contradicted. The heavy elements (earth and water) wouldn't fall in a straight line from top to bottom, the clouds would move westwards and the movement would be so great as to disperse everything, not adapting to Aristotelian cosmology (BASTOS FILHO, 1995).

The ideas of Aristotle and Ptolemy were well accepted during the Renaissance. But for a time they were forgotten. After the decline of the Roman Empire, much of the knowledge of ancient Greece was forgotten. However, the scientific ideas of that time were rediscovered in Europe from the 12th century onwards (KEARNEY, 1970).

The theory that all the stars revolved around the Earth was not accepted by all the Greeks. The Greek Aristarchus of Samos (estimated 301 BC to 230 BC), for example, defended the idea that all the stars, including the Earth, revolved around the Sun. He also considered the Earth's rotational movement. But his ideas were not well accepted at the time, and one of the arguments used at the time against the Earth's movement was that when a person threw an object at

If the Earth were moving around its axis, the person would be moving along with the star, and the body being thrown would fall a certain distance from the person.

In the 14th century, the French bishop of Lisieux called Nicholas Oresme admitted the possibility of the Earth rotating around its own axis on a daily basis. In the following century, the cardinal and philosopher Nicholas of Cusa again discussed the possibility of such a movement. Neither of them were professional astronomers, and they didn't "build" a world system that could "clash" with Aristotelian Cosmology and Ptolemaic Astronomy (SMITH, 1973). The idea of the Earth's rotation was once again discussed by the Polish monk Nicolaus Copernicus (1473-1543). In the Copernican theory, the Moon also went round the Earth and the Earth revolved around the Sun. These statements were different from the Aristotelian-Ptolemaic theory. The movement of the Earth around its axis was the explanation Copernicus used to clarify the apparent movement criticised by the defenders of the geocentric theory. Below, we highlight some of what Copernicus said about this and his system:

> In fact, the main arguments by which physicists try to substantiate the Earth's immobility are based above all on appearances; and here they are the first to fall apart, because we consider the Earth's immobility to be due to that very appearance (COPÉRNICO, 2003, p. 119).

Copernicus didn't publish his ideas for almost four decades. Commentariolus, for example, was written in the 1510s, but was only edited and published in 1878. However, he wasn't very happy with it. Copernicus realised that it was still necessary to formulate a theory explaining planetary movements . Approximately 20 years after Commentariolus was written, it managed to reach Rome and even Pope Clement VII. However, there were no objections to the theory or the author, not even from the Pope. So much so that the cardinal-archbishop of Capua, Nicholas Schonberg, invited Copernicus to publish a book about his discoveries, promising to make copies of the work himself. Copernicus chose not to accept the cardinal's invitation, just as he didn't accept advice from friends who said that a publication on his ideas was a duty not only to the philosophy of nature but also to humanity. Copernicus feared a scandal and a hostile reaction (KOYRÉ, 1960).

He was even harshly criticised by the Protestants Martin Luther and John Calvin. The former went so far as to say:

> **"The new astrologer who wants to prove that the Earth moves and revolves around itself." He went on to say: "The madman wants to turn the whole art of astronomy upside down". He added: "As Holy Scripture tells us,** Joshua commanded the Sun to stop and not the **Earth" (apud SMITH, 1973, p. 97).**

Calvin, on the other hand, said: "...it was determined that the world could not move", and continues with the following question: "Who will dare to place the authority of Copernicus above that of God?" (apud SMITH, 1973, p. 97).

The manuscript of The Revolutions of the Celestial Orbs, Copernicus' best known and most important book, was finished around 1530. Over the years, occasional corrections were made, but Copernicus didn't publish it because he feared exposure to ridicule. Copernicus had no idea of the value of his production. It wasn't until 1542 that his work was put into print. The following year, the book was published (SMITH, 1973). In this book, he tries to explain why a person on Earth can perceive apparent movements:

> ...when a ship is sailing smoothly, everything outside it seems to the sailors to be moving by the reflection of that movement and, on the other hand, they think they are immobile with all the objects next to them.
> Naturally, the same thing happens with the movement of the Earth, so that the whole Universe seems to rotate (COPERNICO, 1984, p. 40 and 41).

In the same book, he illustrated what his proposed heliocentric theory looked like. See below:

Fig. 03 - Copernican system
Source: COPÉRNICO, 1984.

According to Bastos Filho (1995), Copernicus opposed Ptolemy by saying that instead of the whole sky revolving around the Earth, it was simpler for the Earth to revolve around its axis. Based on the formula known today "$V = (2\pi R/T)$", as the size of the universe was greater than the size of the Earth, the speed that the former would have to make in 24 hours would be extremely high, leading to infinity, violating Aristotle's own physics as well as his cosmology. Therefore, it is more reasonable for the Earth to move around itself than the entire universe. See below what Copernicus said in his book The Revolutions of the Celestial Orbs about the above:

> But why doesn't the same question arise even more intensely about the Universe, whose movement must be all the faster because the sky is bigger than the earth? Or did the sky become immense because it was deflected from its centre by a movement of indescribable force and ended up falling too, if it stopped? Certainly, if this reasoning were reasonable, the greatness of heaven would also rise to infinity. Indeed, the higher it is raised by the force of its movement, the faster this movement will be due to the continuous increase in the circumference it has to travel in the period of 24 hours (COPERNICO, 1984, p. 39).

After Copernicus' book The Revolutions of the Celestial Orbs was published, a few generations passed without giving any relevance to the innovative ideas contained in this work. The Italian philosopher Giordano Bruno (1548-1600) was one of the first to defend the Copernican theory, showing how it was changing not only the view of the universe, but also of humanity. But it was natural that at the time a scientific idea didn't have so much "power" to make a radical change. Theology and philosophy, both in the medieval period and even in the Renaissance, were the areas of knowledge worthy of credit, while the sciences represented secondary knowledge. Galileo was a key figure in changing the role of science, fighting to show its value in explaining the phenomena of nature (ORTEGA y GASSET, 1989).

Although Giordano Bruno was not an astronomer, physicist or mathematician, nor did he live long enough to enjoy the telescope, he, through a genius intuition, surmised ideas that were contrary to those defended at the time. Like the idea of an infinite Universe, immense and not innumerable, and full of worlds similar to ours, as well as populated by people similar to human beings. Giordano Bruno defended this view throughout Europe, as well as criticising Aristotelianism. Because of this, he was arrested by the Inquisition in 1593 and imprisoned for seven years. In 1600, in Rome, he was excommunicated and burnt at the stake (KOYRÉ, 1960).

Even though Bruno defended the Copernican theory, by proposing an infinite universe, he removed the Sun from the centre. In his view, the solar system is just one of many that exist in the universe. There is no centre (KOYRÉ, 2009).

For Koyré (1960), the Catholic Church did not consider Copernican ideas to be dangerous before Giordano Bruno. One of the first to speak out against Copernicus was Father Clavius, who criticised the physical absurdities, as well as the theory's inconsistency with many passages in the Bible. Even so, he went so far as to praise Copernicus, seeing him as a second Ptolemy.

Regarding Copernicus' arguments against appearances, Galileo was a great admirer of the Polish astronomer's thinking because he contradicted in favour of reason what the senses showed him in his observations of the movements of the stars. Take a look at what Galileo said in the Third Journey of the book Dialogue in a famous speech by the character Salviati:

> [...]; but those experiments, which openly contradict the annual movement, are of such apparent aversion that [...] I can find no limit to my admiration of how, in

> Aristarchus and Copernicus, reason could do such violence to the senses, that against them it became sovereign of its credulity (GALILEI, 2004, p. 413).

In this text, the Copernican theory will be discussed in the same way as the heliocentric theory. In fact, Copernicus proposed a system in which the stars revolved around the Sun, but the Sun was not exactly at the centre of the whole. He also argued that this star was stationary. Therefore, heliocentrism is not the most appropriate name for his theory, but heliostatic (COHEN, 1967).

With regard to the arguments used by the Aristotelians against the rotation of the Earth, which was defended by Copernicus, as mentioned above, Shea shows some of them in a very objective way:

> ...a stone thrown from a tower would not touch the ground at its base but at a certain distance to the west; cannonballs fired to the west would reach further than those fired to the east; a powerful wind from the east would blow constantly; the centrifugal force of rotation would blow trees and houses into the air (SHEA, 1983, p. 156).

Galileo's most important role in the struggle to get people to accept the Copernican theory was to propose a physics that was contrary to that of the Peripatetics, going against the sensible appearances that led people to misunderstand the movement of the Earth. The ideas of inertia and the principle of relativity, which Galileo contributed to developing, were the arguments that this philosopher used to show that the Earth could have a rotational movement without the problems that the Aristotelians (mentioned above) used against this idea that was contrary to the stability of this planet (GEYMONAT, 1997). This is followed by Galileo's explanation of the aforementioned information, which can be found in the second day of the Dialogue in the speech of the character Salviati:

> Lock yourself and a friend in the largest compartment under the deck of a large ship, and make sure that there are flies, butterflies and similar flying animals there; also place a large container of water with small fish in it; hang a bucket over it, and drop by drop the water into another narrow-mouthed container underneath: And when the ship is at rest, diligently observe how those little flying animals go with equal speed to all parts of the environment; you will see the fish swimming indifferently everywhere; the falling drops all entering the vessel placed below; and you, throwing something to your friend, should not throw with more force to this part than to that part, when the distances are equal; and by jumping, as they say, with your feet together, you will cross equal spaces everywhere. Make sure you have all these things diligently, even though there is no doubt that while the ship is stationary things should happen like this, and make the ship move with as much speed as you

> wish because (as long as the movement is uniform and not fluctuating from here to there) you will not recognise the slightest change in all the effects mentioned, nor will you be able to understand from any of them whether the ship is moving or stationary: By jumping, you would be travelling the same distance on the deck as before, nor would you be jumping more towards the stern than towards the bow, because the ship is moving very quickly, even if, during the time you are in the air, the underlying deck slides in the opposite direction to your jump; and if you throw something at your companion, you won't have to throw it any harder to reach him, if he's in the bow and you're in the stern, than if you were placed the other way round; and the drops will continue to fall as before into the lower container, without any of them falling towards the stern, even if, while the drop is in the air, the ship is sailing many metres; the fish in its water will swim without any greater effort to either the preceding or the subsequent part of the vessel, and will just as easily reach the food placed anywhere on the edge of the container; and finally the butterflies and flies will continue their flights indifferently to all parts, and it will never happen that they concentrate on the part addressed to the stern, as if they were tired of following the swift course of the ship, from which they would be separated by remaining in the air for a long time; And if you were to burn a tear of incense and produce a little smoke, you would see that it rises upwards and like a little cloud it remains there, moving indifferently no more towards this part than towards that part. And the reason for all this correspondence of effects is that the movement of the ship is common to all the things contained in it and also in the air, which is why I suggested that it should be under the cover of the ship (GALILEI, 2004, p. 268-269).

Another system was proposed after Copernicus. The Danish astronomer Tycho Brahe (1546-1601) organised a universe in which the stars revolved around the Sun and all of them went around the immobile Earth, which was at the centre of this system (fig. 04). It was a kind of geo-heliocentrism. According to Santillana (1960), this system, in the view of some people at the time, saved official philosophy and Sacred Scripture. Galileo did not consider Brahe's idea as a third system to be discussed.

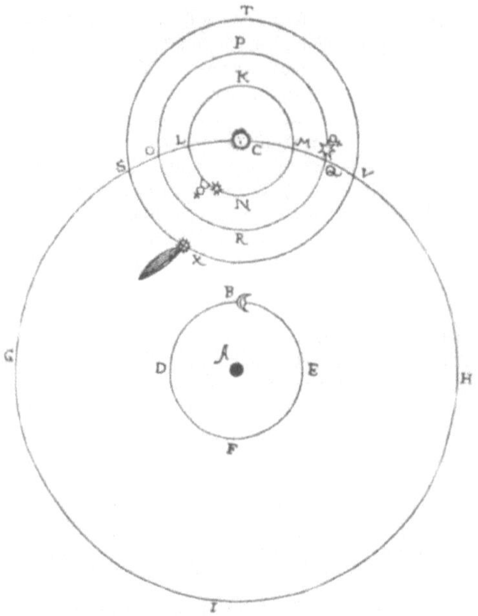

Fig. 04 - World system idealised by Tycho Brahe
Source: http://www.on.br/ead_2013/site/conteudo/cap7-historia/astronomia-renascenca/tycho-brahe/images/tycho2.jpg

In 1604, an event caught Galileo's attention: a new star appeared in the sky (it was a supernova) and it was above the Moon, in the place that the Aristotelians said was incorruptible, a perfect world. This appearance went against this idea. But people who believed in Copernicus' theory were unable to relate it to heliocentrism (FISCHER, 1986).

In the middle of 1609, Galileo learnt of the existence in Holland of an optical instrument that, when viewed through, distant objects could be seen from closer up. Galileo became interested in this artefact and built one. But his initial intention was to make the telescope for use by the military. This interest was due to the possibility of getting a pay rise (RESTON Jr, 1995). For years, Galileo tried to increase his salary, hoping to earn enough money to cover his expenses and debts so that he wouldn't have to give private lessons, since this activity reduced his research time. Galileo's situation became more complicated after the death of his father Vincentio Galilei in 1591, when he had to take care of almost the entire family. He had to pay the wedding dowries of his two sisters Virginia and Livia, he helped his brother Michelangelo financially on several occasions, he looked after his mother and he also had three children (Virginia - 1600; Livia - 1601; Vincenzio - 1606). As mentioned above, the telescope he built was the possibility of

Galileo was able to increase his salary (GEYMONAT, 1997).

Galileo showed his telescope to the Venetian authorities and they liked and approved of his instrument, as they saw the great advantage they would have in a war, as they would be able to observe potential enemies from afar and thus anticipate their actions. Galileo managed to double his salary, as well as earning an immediate bonus. His book Messenger to the Stars describes the construction of his telescope and the moment he learnt of its existence:

> About ten months ago the news reached us that a certain Belgian had produced a **spyglass** with which visible objects, although very far from the observer's eye, could be discerned as clearly as if they were close by; about this effect, which was indeed admirable, some experiments were told, in which some had faith while others denied it. The fact was confirmed to me a few days later in a letter from the noble Frenchman Jacques Badouvert, from Paris, which was the reason that led me to dedicate myself fully to the search for the reasons, no less than to the search for the means by which I could invent a similar instrument, which I achieved shortly afterwards, based on the doctrine of refractions. Firstly, I prepared a lead tube at the ends of which I fitted two glass lenses, both flat on one side while the other side was convex on one lens and concave on the other. Thus, with my eye on the concave, I saw objects so large and close that they seemed three times closer and nine times larger than when observed only with natural vision. Later, I made a better one, which represented the objects more than 60 times larger. Finally, with neither expense nor fatigue, I managed to make an instrument so excellent that the things seen with it seemed almost a thousand times larger and more than 30 times closer than when observed with the natural faculty (GALILEI, 2009b, p. 33).

When Galileo started building his own telescopes, he didn't know, like all the other philosophers, the scientific knowledge of this optical instrument. Although he knew that with it he would be able to see distant objects more closely, he couldn't explain scientifically why this was the case. Therefore, this invention was not scientific, but technical. Kepler's work Dioptrica gave us an idea of how the telescope worked scientifically. However, the most important thing was how Galileo used the telescope to unveil knowledge of the sky that contributed to a new vision of astronomy (BANFI, 1949). Although he wrote in his book that he had based himself on the doctrines of refractions, as described above, he did not have an exact scientific notion of the subject.

With his telescope, in 1609 and 1610, Galileo observed craters and mountains on the Moon and four satellites of Jupiter. These discoveries were published in his work entitled Messenger to the Stars. In this book, he drew images of the Moon and the four satellites of Jupiter from his observations made with the telescope on different days. Below are these drawings and what he said about these discoveries:

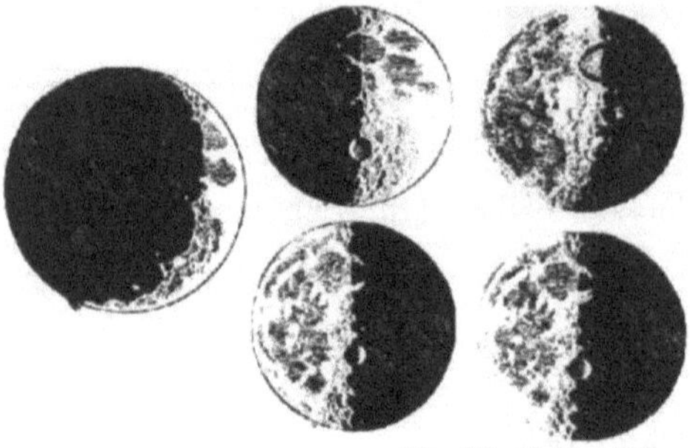

Fig. 05 - Galileo's drawings of the Moon
Source: GALILEI, 2009b.

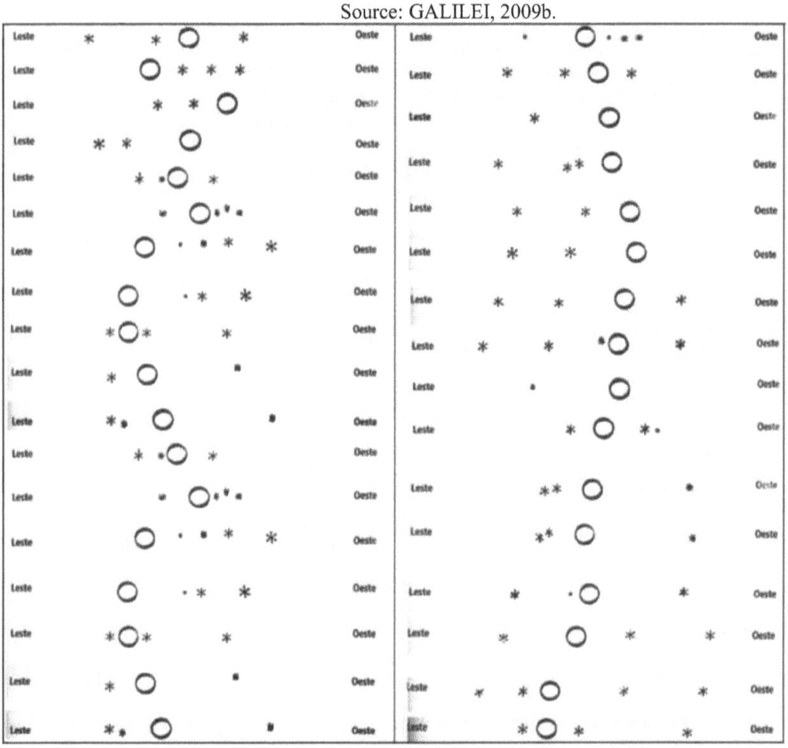

Fig. 06 - Galileo's drawing of Jupiter and its four satellites.
Source: GALILEI, 2009b.

> Here, then, are four stars[4] ... which, with disparate movements among themselves, perform their courses and orbits around the star Jupiter, the noblest of them all, as unanimously agreed, great revolutions around the centre of the world, that is, around the Sun (GALILEI, 2009b, p. 28 and 29).

> ...after careful and repeated inspections, we have deduced the opinion, which we hold firmly, that the surface of the Moon and the other celestial bodies is not, in fact, smooth, uniform and of the most exact sphericity as has been taught by a numerous court of philosophers, but that, on the contrary, it is uneven, rough and full of cavities and prominences, not unlike the face of the Earth itself, which presents, here and there, the crests of mountains and the abysses of valleys (GALILEI, 2009b, p. 36).

Galileo realised that the Moon had craters and mountains. These observations went against the idea of the "absolute perfection" of the supralunar stars defended by Aristotle. If these stars were perfect, and this required them to be perfectly spherical, then the Moon could not have these "imperfections". Many people did not accept what had been observed. But Father Clavio, in order to reconcile what had been discovered on the Moon with the ancient Aristotelian idea of the perfection of this star, invented that these craters and mountains were filled with a transparent crystalline substance, leaving it perfectly spherical and smooth (GEYMONAT, 1997). Galileo sarcastically responded to this defence: "Imagination really is beautiful... the only thing it lacks is that it is neither demonstrated nor demonstrable" (Apud GEYMONAT, 1997, p. 69). The philosopher made it clear that since Clavio's idea was neither demonstrable nor proven, it could not be considered "true" within the realm of science, being a beautiful fantasy created by this priest (GEYMONAT, 1997).

Going against Aristotle's idea of perfection was nothing new. In 1572, for example, the Danish astronomer Tycho Brahe observed a star "lighting up" in the sky, and made calculations to find out where this phenomenon had occurred. His measurements showed that it had appeared after the planets, in the sky of the fixed stars. In other words, it took place in the supralunar world, going against the immutability of the sky as defended by Aristotle. Due to Brahe's measurements, when

[4] These stars that Galileo mentions are what we call Jupiter's natural satellites today.

a new star appeared in 1604, the debates intensified. People who defended Aristotle's thoughts sought arguments against Brahe's calculations and ideas. One of these defences was that the star was ancient and it was only then that it was possible to see it. Another was that it had been created, adding a term that did not contradict incorruptibility. And finally, they said that either the calculation was wrong or, if it was accurate, it wasn't a star but the reflection of the sun's rays (BANFI, 1949).

The observation of Jupiter's four "moons" was a discovery that helped the heliocentric theory, since the confirmation of the existence of these satellites circling a planet contributed to Copernicus' assertion that the Moon did the same around the Earth, while the latter revolved around the Sun.

The four satellites that Galileo observed circling Jupiter were called "Median planets". He didn't honour the Mediei family unintentionally. The strongest funding he could have had at the time was from the church and/or the ruling houses. As Galileo knew that his discoveries were going to be criticised in the scientific world, having the support of this family would benefit him (GEYMONAT, 1997).

Galileo wasn't the first to point the telescope towards the sky, but he was the first to give due importance to the observations made using this optical instrument, when he realised that these discoveries (not only those mentioned above, but also others that will be discussed below) were not in line with the Aristotelian-Ptolemaic geocentric theory, but rather fitted in with Copernicus' heliocentric theory (GEYMONAT, 1997).

During this period, more precisely in 1609, the German Johannes Kepler (15711630) published a book entitled Astronomia Nova. In this work, one of the ideas he defended was that the planets orbited the Sun in an elliptical trajectory with the star occupying one of the foci[5] . This concept was not widely accepted, even by supporters of the Copernican theory. Galileo, for example, did not agree, as he argued that the trajectory of the planets around the Sun was circular (COHEN, 1967).

Another observation made by Galileo was the illumination of the planet Venus. He realised that this planet had phases like those of the Moon, and this detail was

---

[5] This idea is known as Kepler's first law.

important in questioning, once again, a flaw in the theory of geocentrism in relation to the discoveries that were being made. In order to observe the full phase of the planet Venus, this happened in heliocentrism because the Sun could be between the Earth and the Moon.

Venus. In geocentrism, this didn't happen, because Venus was always between the other two stars. Another detail is related to distance, due to the fact that in the full phase of Venus they saw the planet as smaller, and in the new phase as larger. In Copernican theory, this was because Venus was able to stand at different points in relation to the Earth; at times it was close to the Earth and at other times much further away from it, making it possible to see its visible disc smaller at full, and in the phases new, crescent and waning, larger, since in the latter Venus would be closer to the Earth. Galileo realised these size differences in his observations. In the geocentric theory, it wasn't possible to observe such a difference in the size of Venus (COHEN, 1967). See below the drawing made by Galileo of how he saw Venus through his telescope, showing the phases of this planet and the different sizes in these "different" moments of this star.

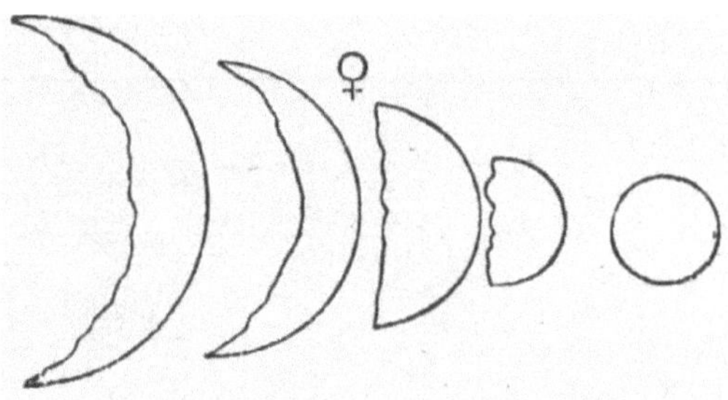

Fig. 07 - Galileo's drawing of Venus.
Source: Book Galileo's **Rehearser**

Another important observation made by Galileo was that of sunspots. This discovery went against the idea of the perfection of the supralunar world defended by Aristotle. According to Fischer (1986), this new fact was more dangerous to Aristotelian cosmology than the craters and mountains on the Moon. As the latter star was on the border between the sublunar and supralunar worlds, it was only a matter of the Moon becoming part of the former. That way, not much would change. But it wouldn't be possible to make that kind of change using the Sun.

The moment that Galileo observed the sunspots and then published this discovery was the beginning of a fight with the astronomer and Jesuit professor Christoph Scheiner, who claimed to have seen the sunspots before Galileo (FISCHER, 1986). According to Shea (1983), Scheiner wrote a letter dated 12 November 1611 to his friend Mark Welser stating that he had observed sunspots seven months earlier. Welser copied this letter, as well as two others he received from Scheiner, and sent them to a number of people, including Galileo. As Scheiner's superiors were concerned about publishing a discovery that could be mistaken, they ordered him to sign the letter with a pseudonym. Galileo returned a letter to Welser, dated 14 May 1612, saying that he had already observed the sunspots eighteen months earlier. However, the question of who observed the sunspots first is a historical detail that we won't go into as it's not part of the purpose of this text.

Some Aristotelians of the time argued that these sunspots were not on the surface of the Sun, and even proposed clouds in front of it. Galileo contested their views and even proposed that the Sun was rotating (SHEA, 1983).

Below is Galileo's drawing of the sunspots:

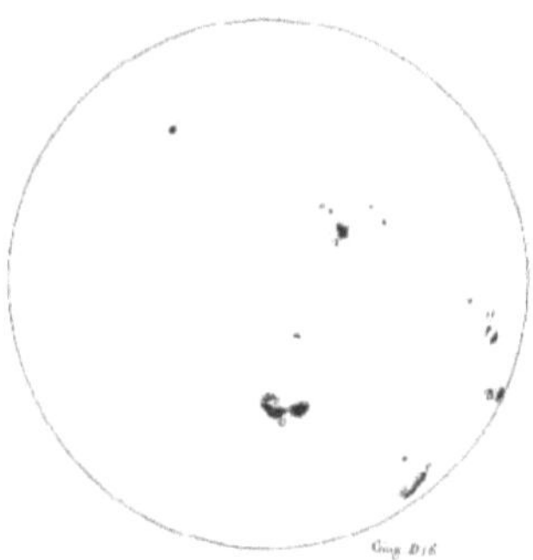

Fig. 08 - Galileo's drawing of sunspots.

Source: Article by MOSCHETTI, 2006

Some scholars at the time argued that the ideas related to the heliocentric theory went against some passages in the Bible. Galileo even commented on this:

> The mobility of the Earth and the stability of the Sun can never be against faith or Sacred Scripture if it is truly proven by philosophers, astronomers and mathematicians, with sensitive experiments, careful observations and necessary demonstrations that it is true in nature. But in this case, if some passages of Scripture

> seem to sound to the contrary, we must say that this is because of the weakness of our intelligence, which has not been able to penetrate the true teaching of Scripture in this particular. This is a common and very correct doctrine, since one truth cannot contradict another truth. Therefore, anyone who wishes to condemn it according to law must first demonstrate that it is false in nature, answering the reasons to the contrary (GALILEI, 2009a, p. 119 and 120).

On 21 December 1613, Galileo wrote a letter to Benedetto Castelli which ended up being used against the philosopher as he began a period of conflict with the Church, as he commented in this letter on the interpretations made of Holy Scripture in relation to astronomical events and knowledge. According to Galileo:

> ...although Scripture cannot err, no less could some of its interpreters and expositors sometimes err in various ways (GALILEI, 2009a, p. 18 and 19).
>
> ...not every saying of Scripture is bound by obligations as severe as every effect of Nature. On the contrary, if Scripture has not refrained from obscuring its main dogmas just by accommodating itself to the capacity of rude and uneducated peoples, even attributing to God himself conditions that are very distant and contrary to his essence, who will want to maintain with certainty that, leaving aside this stance, when speaking, even incidentally, of the Earth, the Sun or any other creature, it has chosen to contain itself with all rigour within the limited and restricted meanings of the words? (GALILEI, 2009a, p. 20).

Galileo also commented on the Joshua passage in the Bible. Note what it says:

> Then Joshua spoke to the Lord on the day that the Lord delivered the Amorites into the hand of the children of Israel, and he said in the presence of the Israelites, "Sun, stand still at Gibeon, and you, moon, in the valley of Aijalon.
>
> And the sun stood still, and the moon stopped, until the people took vengeance on their enemies. Isn't this written in the Book of the Righteous? So the sun stood still in the midst of heaven, and did not hasten to go down for almost a whole day (BIBLE, p. 209).

Note what Galileo said about this passage in his letter:

> ...it is necessary to change the meaning of the words and say that when Scripture says that God stopped the sun, it meant that he stopped the first mover, but that in order to accommodate the capacity of those who, with difficulty, are able to understand the rising and setting of the sun, it said the opposite of what it would have said, speaking to wise men (GALILEI, 2009a, p. 25).

Father Lorini, through Cardinal Sfrondratin, sent a copy of the letter written by Galileo to Castelli on 7 February 1615 to the Holy Office in Rome with the intention of examining it to see if the opinions put forward were heretical and/or went against the teachings of the Church. Lorini even altered some of the words in the copy he sent and thus changed the meaning of some of Galileo's arguments (SANTILLANA, 1960).

When he was informed that he had been denounced, Galileo learnt that they had delivered a copy of his letter to the church, so he made a correct copy of it and sent it to his friend Monsignor Piero Dini, who had a post in the Roman Curia. His friend made more copies and gave them to various personalities in Rome. Cardinal Maffeo Barberini, the future Pope Urban VIII, also sided with Galileo (FISCHER, 1986).

In his letter to Dini, dated 16 February 1615, Galileo explained why he had sent an original copy of the letter he had written to Castelli:

> I don't know where he got a copy of a letter I wrote last year to Father Mathematico of Pisa about quoting the sacred authorities in discussions about nature and in explaining the passage from Joshua, but they're shouting about it and finding many heresies in it, and in short, they've opened up a new field to tear me apart. But I'm beginning to suspect that whoever transcribed it may have inadvertently changed a few words, since I haven't been given even the slightest hint of difficulty by anyone else who has seen the letter. This change, coupled with a bit of a penchant for criticism, can make things appear very different from what I intended. And because some of these priests, in particular the one who spoke, have come here to make, as I think, some other attempt with their copy of my aforementioned letter, it seemed to me not out of place to send Your Most Reverend Lordship a copy of it exactly as I wrote it (GALILEI, 2009a, p. 28).

In this same letter Galileo defended Copernicus and said that the monk's ideas should not be condemned. A month later, he wrote to Dini again and commented on the studies carried out by Copernicus:

> ...as for saving the appearances, Copernicus himself had already done the work and satisfied the astronomers in the customary and accepted manner of Ptolemy. But then, putting on the clothes of a philosopher and considering whether such a constitution of the parts of the universe could really exist **"in the nature of things" and seeing that it didn't, and yet seeming to him** that the problem of the true constitution was worthy of investigation, he gave himself over to the investigation of this constitution. Knowing that if an arrangement of fictitious and untrue parts could satisfy appearances, this would be obtained much more with the true and real arrangement and, at the same time, one would have obtained in philosophy a knowledge as excellent as knowing the true arrangement of the parts of the world. Having found himself, thanks to the observations and studies of many years, very well informed of all the particular accidents observed in the stars, without whose very diligent learning and without whose very agile fixation in the mind it is impossible to arrive at the knowledge of the constitution of the world, he achieved, with repeated studies and long labours, what he then made worthy of admiration by all those who study him with application so that they may profit from his progress

> (GALILEI, 2009a, p. 36).

That same year, Galileo sent a letter to the Grand Duchess mother of Tuscany Cristina of Lorraine. He begins the letter with an account of his discoveries related to observations of the sky:

> A few years ago, as Your Serene Highness well knows, I discovered many peculiarities in the sky that had remained invisible until now. Whether because of their novelty, or because of some of their consequences, which run counter to certain propositions about nature commonly accepted by the schools of philosophers, these discoveries have stirred up a good number of their professors against me; almost as if I, with my own hand, had placed such things in the sky to upset nature and the sciences (GALILEI, 2009a, p. 49 and 50).

In the course of the letter, he talks again about these discoveries, how they were contributing to the acceptance of the Copernican theory, as well as defending it:

> ...In my studies of astronomy and philosophy, I maintain, with regard to the constitution of the parts of the world, that the Sun, without changing its place, remains situated at the centre of the revolutions of the celestial orbs and that the Earth, which revolves on itself, moves around it; moreover, I realise that I am confirming this position, not only with the refutation of the reasons of Ptolemy and Aristotle, but with the presentation of many reasons to the contrary; in particular, some relating to natural effects whose causes perhaps cannot be determined in any other way, and other astronomical reasons derived from many comparisons with new celestial discoveries, which openly refute the Ptolemaic system and admirably agree with and confirm this other position (GALILEI, 2009a, p. 52). 52).
>
> **"...which** (Copernicus' **The *Revolutions of the Celestial Orbs***), then in print, was received by the Holy Church, read and studied throughout the world without the slightest shadow of uneasiness ever being discovered in its doctrine. Now, while it is being discovered how well founded it is on manifest experience and necessary demonstrations, there is no shortage of people who, having never seen such a book, reward its author with the dishonour of having him **declared a heretic" (GALILEI**, 2009a, p. 54 and 55, emphasis added).

At various points in this letter, Galileo quoted some of St Augustine's words, making him an intellectual reference in theology who showed that there was no problem in discovering new ideas through science, as long as they were duly "proven", even if they were contrary to the Holy Scriptures. But in that case, theologians and philosophers would have to reinterpret the passages contrary to the new ideas. Take a look at the words of Saint Augustine that Galileo quoted for this defence:

> The following must be said indubitably: whatever the wise of this world can truly demonstrate about the nature of things, let us show that it is not contrary to our Scriptures; whatever they teach in their books that is contrary to the Holy Scriptures, without any doubt let us believe that it is something completely false, and in whatever way we can, let us also show it; Let us thus keep the faith of our Lord, in whom are hidden all the treasures of wisdom, so that we may neither be seduced by the loquacity of a false philosophy nor be frightened by the superstition of a pretended religion (AUGUSTINE apud GALILEI, 2009a, p. 74). 74).

After Father Lorini denounced Galileo for defending opinions contrary to scripture, and therefore those of the Church, and many people accused this philosopher of the same, Cardinal Bellarmine, in 1616, was ordered to summon Galileo, asking him to renounce his opinions regarding the Copernican theory, not to teach it, not to defend it, and not to submit it to discussion. If he didn't comply, he could be imprisoned. Despite this, his works were not included in the Index, but Copernicus' work, The Revolutions of the Celestial Orbs, was suspended until the "necessary" corrections were made to it (BANFI, 1949). It took around a year from the time the Holy Office opened proceedings against Galileo to the events mentioned above. Galileo was also accused by Lorini of wanting to adapt the text of Sacred Scripture with ideas related to the heliocentric theory (REDONDI, 1990). In a letter that Bellarmine wrote to Father Paolo Antonio Foscarini, dated 12 April 1615, he commented on the new ideas and the Holy Scriptures:

> I say that it seems to me that Your Paternity and Mr Galileo have acted prudently, being content to **speak "by supposition" and not absolutely,** as I have always believed Copernicus to have spoken. Because to say that, assuming that the Earth is moving and the Sun is stationary, all appearances are saved better than with the assertion of eccentrics and epicycles, is very well said and there is no danger whatsoever. That's enough for the mathematician. But to want to affirm that the Sun really is in the centre of the world and only revolves on itself without running from East to West, and that the Earth is in the 3rd heaven and revolves with great speed around the Sun, is a very dangerous thing, not only to irritate all scholastic philosophers and theologians, but also to harm the Holy Faith by making the Sacred Scriptures false (BELARMINO, 2009, p. 131 and 132) .[6]
>
> I say that if there were a true demonstration that the Sun is in the centre of the world and the Earth in the third heaven, and that the Sun does not go around the Earth but the Earth goes around the Sun, then we would have to proceed very carefully in explaining the Scriptures that seem to be contrary, and say rather that we don't understand them than that what is demonstrated is false (BELARMINO, 2009, p.

[6] Extract from the book: GALILEI, G. Ciência e Fé: Cartas de Galileu sobre o Acordo do Sistema Copernicano com a Bíblia. (Translation and Org. Carlos Arthur R. do Nascimento) São Paulo: Editora UNESP, 2ª ed.

132 and 133) .[7]

Galileo only became hopeful again in 1623, to the point of believing that a new era was beginning and that he would now have more space to publish his ideas related to astronomy. He began to think this way because Cardinal Maffeo Barberini had been elected the new Pontiff, and was now called Pope Urban VIII. As well as being highly intelligent, he had a sensibility that could contribute to the cultural demands of the time. Urban VIII went so far as to praise Galileo not only for his writing skills, but for being a good Catholic: "We observe in him not only literary distinction, but also a love of religion worthy of Papal esteem" (PAPA URBANO VIII apud RESTON Jr, 1995, p. 255).

In 1623, Galileo published The Assayer. According to Redondi (1990), the problem that Galileo had with the Church after publishing the Dialogue, which we will discuss below, did not begin at that time, but mainly since the publication of The Assayer. In this book, Redondi also states that the philosopher implicitly and explicitly defended Democritus' doctrine of atomism, and this idea did not fit in with the Christian faith of the Eucharist, which was considered one of the most important of the sacraments of the Christian religion. Therefore, the ideas defended by Galileo, in this last book mentioned, offended the beliefs of the Church. But scientifically, there was no reasonable idea at the time to explain how bread and wine became the body and blood of Christ. Shea (1983) points out that when Galileo declared the subjectivity of the qualities of colour, smell and taste in the transformation mentioned above, he endangered the Christian doctrine of transubstantiation. And it was this doctrine that explained this change of bread and wine into the body and blood of Christ, which was not a field for science, but for philosophy and theology.

In 1624, when Maffeo Barberini was already Pope Urban VIII, Galileo travelled to Rome and was received by him six times over six weeks. The philosopher tried to convince him of the importance of discussing Copernicus' ideas, but most of the times he asked him about it, he was unsuccessful. However, he left with the feeling

---

[7] Idem.

that was free to write about the heliocentric theory (SHEA, 1983).

The written part of the book Dialogue on the Two Maximum Systems of the Ptolemaic and Copernican Worlds was completed in 1629. From that moment on, the problems related to the publication of this work began, since in it Galileo addressed Copernicus' theory not in a hypothetical way as the Pope had wished. Castelli asked Galileo to go to Rome to deal with the formalities surrounding the book. When he arrived in Rome, the first revision of the work was made by Father Riccardi, who asked Father Rafael Visconti to analyse the scientific part of the book. After these revisions, Galileo only had to make a few changes. After leaving Rome, Castelli suggested that Galileo speed up the publication process in Florence, since if he did it there first, he would probably have greater difficulties in completing the task and might not even succeed. But Galileo needed Rome's consent, and it was taking a long time to get it. In the end, another revision of the work was made, but it was analysed by a priest from Florence, and once again Galileo had no problems with this, as minor corrections were requested. However, the proem and the final part of the work were to be sent from Rome. However, this did not prevent the book from being published in 1632 (BANFI, 1949).

According to Santillana (1960), with regard to the name of this book, Galileo initially wanted it to be called Del Fluxo y Refluxo del Mar. However, after arriving in Rome in 1630, Urban VIII suggested that he change the title of the work to Sobre los Dos Principales Sistemas del Mundo.

In Dialogue, Galileo created three characters. The first was called Salviati, a convinced Copernican who used Galileo's own arguments. The author used him to defend his ideas. The second was Simplicio, an Aristotelian who defended the ideas of the geocentric theory of Ptolemy and Aristotle, and the latter's physics. The last was Sagredo, an intelligent person who was easily convinced by Salviati's arguments. According to Santillana (1960), Sagredo was a Venetian nobleman who was attentive to the development of science, open to discussions without compromising on more theoretical problems. Simplicio was an Aristotelian and stubborn university professor who lost his temper when he felt cornered in debates, but managed to regain his composure.

Koyré (2009) highlights the way Galileo wrote this book, his proposal and the dialogue between the characters, saying:

> The **Dialogue on the World's Two Greatest Systems** aims to explain two rival astronomical **systems**. However, in reality, it is not an astronomy book, nor even a physics book. It is, above all, a book of criticism; a work of polemic and combat; it is at the same time a pedagogical, philosophical work; and, finally, a book of history: <la historia del pensamiento de Galileo>.
> A work of polemic and combat; this is what partly determines the literary structure of the **Dialogue**: Galileo arms his war machine against traditional science and philosophy. However, if the **Dialogue** is directed against the Aristotelian tradition, it is not intended - or rather - for **its defenders... it** is **intended for the <honoured> reader; that** is why it is not written in Latin - the erudite language of universities and schools - but in vulgar language, in Italian, which is the language of the court and the bourgeoisie (KOYRÉ, 2009, p. 200 and 201).

He goes on to say:

> A <pedagogical> work. Pues no trata só de convencer, persuadir y probar; también - y quizá sobretudo - se trata de poner, poco a poco, al lector honrado en condiciones de ser persuadidos y convencido; en condiciones de comprender la demostración y admitir la prueba. And this requires a double task of destruction and education: the destruction of prejudices and traditional mental habits and of common sense; the creation, instead, of new habits, of a new aptitude for reasoning (KOYRÉ, 2009, p. 201).

Despite Galileo's great contributions to the acceptance of the Copernican theory, in his book Dialogue, when he tried to defend the Earth's rotational movement with his explanation of the phenomenon of the tides, he made a mistake in his idea. For Galileo, because at some moments (as at point D in fig. 09) the Earth's rotation and translation around the Sun were in the same direction and at others (as at point F in fig. 09) in different directions, there was a difference in the planet's speed due to acceleration and retardation, causing the tides (MARTINS, 1994).

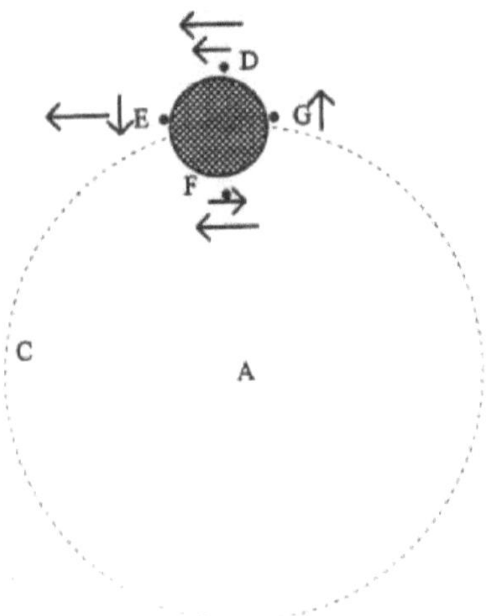

Fig. 09 - Vectors Indicating the Earth's Rotational and Translational Movements.
Source: Article by MARTINS, 1994

Scheiner and Grassi tried to convince Urban VIII that Simplicio was the Pope himself, which made him angry, since the character was a fool, a simpleton, a fool, the target of the mockery. Many of the opinions used by this character in the book, which were ridiculed, were similar or the same as those of the Pontiff himself, making him even more upset. Galileo even asked some influential people, such as the Pope's own brother, to take the idea out of Urban VIII's head. In addition, some Spanish cardinals were accusing the Pontiff of not taking firm decisions against people who were being accused of heresy. And the problem with Galileo seemed to be a great opportunity for the Pope to show how tough he could be with these people (RESTON Jr, 1995).

According to Fischer (1986), in 1616, Pope Urban VIII, who at the time was Cardinal Maffeo Barberini, supported Galileo in the trial that year, but became the philosopher's enemy after the problems that arose following the publication of the book Dialogue.

The Church organised a commission to analyse Galileo's book. After five sessions, held over a period of one month, the commission sent the Pope three problems it found with the work. The first was that Galileo had transgressed the orders by treating Copernican ideas as hypothetical, defending the movement of the Earth and the immobility of the Sun. The second was that, in the eyes of Church science, he wrongly described the phenomenon of the tides with the stability of the Sun and the movement of the Earth. And lastly, the philosopher cheated by bypassing the order issued by the Holy Office in 1616. Even with these problems, the commission came to the conclusion that the work only needed a few corrections (SANTILLANA, 1960). Even though the commission came to this conclusion, the Pope took other measures.

Cardinal Antonio Barberini entrusted the inquisitor of Florence with the task of informing Galileo that he had been ordered to move to Rome; he would be at the disposal of the commissary general of the Holy Office. The philosopher, who was not in very good health and was of advanced age, asked several powerful people at the time to convince the Church authorities that he was in no condition to make the long journey. Galileo managed to postpone his trip to Rome for a few months, but after a while, Barberini wrote a letter to the inquisitor of Florence demanding that the philosopher go to Rome as soon as possible, and would not accept any more excuses for not travelling there (GEYMONAT, 1997). Take a look at the letter below:

> This Congregation of the Holy Office has very much misunderstood that Galileo Galilei did not promptly obey the order to come to Rome; and he must not excuse his disobedience with the season, for it is his own fault that he has come to these times; and he does a great deal of harm by trying to disguise himself by pretending to be ill, for the Holiness of Our Lord and of these Most Eminent Gentlemen of Mine does not want to tolerate these pretences in any way. In the meantime, may Your Reverence tell him that if he does not obey, a Commissioner with a doctor will soon be sent from here to take him and bring him to the prison of this Supreme Tribunal, bound in irons, since it is clear that he has abused the kindness of this Congregation, for which he will be condemned to all the expenses incurred for this purpose. Your Excellency must carry out what has been imposed on him and give him notice (BARBERINI apud GEYMONAT, 1997, p. 193).

According to Geymonat (1997), Urbano VIII favoured this group on many occasions because the Pope received support from the French cardinals. This stance influenced other groups to oppose him, creating a political crisis in 1632, which also influenced the Pope's stance towards Galileo. A critical moment was during the Consistory held that same year, when Cardinal Gaspare Borgia, the Spanish ambassador, openly attacked Urban VIII, accusing him of not being tough enough on heretics to protect many of them. The Pope realised the delicacy of this accusation in relation to his power, and found many opponents everywhere, as some intended to poison him. As a result, he began to try to make amends. Faced with this circumstance, he probably listened to many of Galileo's opponents, strengthening the idea that the Dialogue was made to denigrate his image. In this way, it would be prudent to take revenge on the philosopher, also showing the Pontiff's opponents that he, Urban VIII, was defending Christian interests. This situation infers that the punishment of Galileo was not only carried out for the scientific content of the book, but was also an opportunistic move by the Pope to defend his dignity and authority in the face of his position in the Vatican.

According to Redondi (1990), the Florentine ambassador Niccolini, in one of his audiences with the Pope, went to deliver a note of protest from the Florentine government against the commission that had been set up to analyse Galileo's Dialogue. Urban VIII, who was impatient with the process, said that: "Galileo had osado entrar donde no debía y además en las materias más graves y peligrosas que en estos tiempos se pudieran suscitar" (URBANO VIII apud REDONDI, 1990, p. 302). Note what Niccolini said about this moment:

> I contested that Mr Galileo had not printed the work without the Vatican's approval. The Pope contested, with equal fury, that he and Ciámpoli had warned him, especially Ciámpoli, who had expressed to him that Galileo would be guided in everything by the papal orders and that everything would turn out well; and that was all he had known, without ever having seen or read the book. He was bitter about Ciámpoli and Father Riccardi, although he said that the latter had been forewarned in his turn, because with beautiful words he had first taken the licence and then the permission to print in Florence, without following the instructions given to the Inquisitor, and then also putting the name of Riccardi, which had nothing to do with licences outside Rome. And here I took the floor again to say that I knew about the appointment of a special committee and, since it could very well happen that there were people on it who were ill-disposed (as is the case) against Galileo, I respectfully begged him to give them the opportunity to justify themselves. To which His Holiness replied that, in matters of the Holy Office, **one never does anything but pass sentence and then summon him to retract. I contested: "4N0** does it seem to Your Holiness that Mr Galileo should have been informed beforehand of the difficulties, oppositions and censures that his work is facing, and what it is that **displeases the Holy Office?" And his reply was violent: "The Holy Office,** we assure you, sir, does not proceed in this way, nor does it follow this path, nor does it provide information in advance. It is not customary. On the other hand, he knows well what the difficulties are, if he wants to know them; because we have discussed them with him and he **knew them for us." And since I thought that the book had been dedicated to the** Serenísimo Maestro, that it was the work of one of his servants and that I humbly hoped that some consideration would be shown, I said that in these things that can cause serious harm to religion, of the worst that has ever been invented, His Highness must agree to punish them, because he is a Christian prince and, consequently, I must write to him so that he doesn't get mixed up in it if he wants to leave with honour

(NICCOLINI apud SANTILLANA, 1960, p. 170 e 171).

Niccoline was the one who informed Galileo that he had to appear before the Holy Office. At the first meeting, on 12 April 1633, he was initially asked if he knew why he was there. The philosopher replied that he thought he was there because of the book Dialogue. When they showed him the work, Galileo recognised it (SANTILLANA, 1960). The philosopher was asked about the events of the 1616 episode. Disagreements began about this moment. The notification that Galileo said he had received from Cardinal Bellarmine did not tally with the document shown by the commissioner. The latter said that the movement of the Earth and the immobility of the Sun should not be taught in any way. Galileo then reported that at no time was he informed of this detail, but that he was told that he could neither support nor defend Copernicus' ideas, which went against Sacred Scripture. He argued that at no point in the book Dialogue did he defend or support Copernicus' theory; on the contrary, the ideas of heliocentrism were shown to be neither valid nor conclusive. These last words were clearly lies used by Galileo to avoid committing himself. Even so, these statements made by the philosopher led three theologians to examine his work. After their evaluation, they were unanimous in saying that the book went against the decree of 1616. Two of them reported the suspicion that Galileo continued to defend ideas related to the Copernican theory, while the other was emphatic in saying that the entire work was a defence of this theory (BANFI, 1949).

During the trial, according to Fischer (1986), Galileo was not imprisoned, as was customary for most people who found themselves in the same position as him. He stayed in a more comfortable room in the inquisitorial palace. This only happened due to the great efforts of the Grand Duke of Tuscany and his ambassador in Rome.

After the first meeting, they tried to convince Galileo that he should confess his defence of heliocentrism in his book. At the second meeting, they asked Galileo if he had anything to say in order to get him to admit his "mistakes". Galileo said that when he reread the book, after some time without reading it, he had the feeling that he was dealing with a work by another author. He said that although he had written the book without pretending to defend Copernican ideas, in some passages the arguments used could lead a more ignorant reader to believe in the heliocentric theory. He then confessed to having boasted with his words (SANTILLANA, 1960).

According to Santillana (1960), Niccolini managed to have an audience with Urban VIII before Galileo's last encounter with the Inquisition. But the conversation was not very encouraging. The Pontiff informed him that Galileo would be forbidden to give his opinion on the heliocentric theory, since Copernican ideas were contrary to the Holy Scriptures. He anticipated that Galileo would be imprisoned for not having obeyed the 1616 document. Niccolini informed Galileo of what the Pope had told him.

In the last interrogation by the Inquisition, Galileo was asked if he held or had ever held the idea of the immobility of the Sun and the movement of the Earth. Galileo said that he had come to believe that the ideas of Ptolemy and Copernicus were worthy of discussion and that either could be true. But that the wisdom of the authorities had shown him that only Ptolemy's theory was true. Galileo was then asked how he discussed these issues in his book. Galileo replied by saying that he wrote the book trying to show the arguments that defended both theories respectively, without pretending to prove either of them. On the contrary, the two did not have the decisive strength to reach a conclusion. And that, because of this, he had to resort to the teaching of the authorities (SANTILLANA, 1960).

After these inquisitions, Galileo was taken to the Dominican convent of Santa Maria de Minerva. In front of the congregation of the Holy Office, which was present at the time, the sentence was read that banned the work Dialogue, condemned the author to private imprisonment and certain penances (for three years he had to recite the seven Psalms once a week). After hearing his sentence, Galileo, kneeling, read his abjuration (GEYMONAT, 1997):

> I, Galileo, son of Vincentio Galilei of Florence, at the age of 70, personally constituted in judgement and kneeling before you, Most Eminent and Most Reverend Cardinals, in the entire Christian republic general inquisitors against heretical depravity, holding before my eyes the sacrosanct Gospels, which I touch with my own hands, swear that I have always believed, believe now and with God's help will believe in the future, all that the Holy Catholic and Apostolic Church affirms, preaches and teaches. But because of this Holy Office. Because I, after having been legally summoned, had to totally abandon the false opinion that the Sun is the centre of the world and does not move, and that I could not support, defend or teach the said doctrine in any way, either orally or in writing, and after having been notified

that the said doctrine is contrary to Sacred Scripture, written and printed a book in which I treat of the same doctrine already condemned and adduce reasons very effectively in its favour, without presenting any solution, I was judged vehemently suspected of heresy, that is, of having affirmed and believed that the Sun is the centre of the universe and immovable and the Earth is not the centre and moves. Therefore, wishing to remove from the mind of Your Eminence and all the Christian faithful this vehement suspicion, rightly conceived, with a sincere heart and unfeigned faith, I abjure, curse and detest the aforementioned errors and heresies and, in general, any and all errors, heresies and sects contrary to the Holy Church; and I swear that in the future I will never again say, either orally or in writing, such things as may subject me to similar suspicions; but if I meet any heretic or suspected heretic, I will denounce him to this Holy Office or to the Inquisitor or Ordinary of the place where I am. I also swear and promise to fulfil and fully observe all the penances that have been or will be imposed on me by this Holy Office; and if I break any of my promises and oaths, which God forbid, I will submit to all the penalties and punishments that are contained in the Sacred Canons and other general and special constitutions imposed and promulgated against such offenders. So help me God and these holy Gospels of yours, which I touch with my own hands (GALILEO apud BANFI, 1949, p. 232 and 233).

According to Reston Jr (1995), since the Church itself had authorised the publication of the work Diálogo, and was now condemning it for what was written in it, action was taken against people directly and indirectly involved with The Galileo Affair. Father Riccardo, who had authorised the publication of the book, was dismissed from his post as master of the sacred palace. The inquisitor of Florence, who allowed the book to be published in that city, was reprimanded. Monsignor Ciampoli, Galileo's agent and an intimate of the Pope, was exiled from Rome and spent the rest of his life in minor ecclesiastical posts. Father Castelli, fearing punishment, left Rome, only to return months later.

Within the Catholic Church there were several people with different thoughts, since three cardinals (out of ten responsible) refused to sign the judgement against Galileo. Possibly, they only took this attitude because they believed in the philosopher's faith (DRAKE, 1981).

It wasn't until more than three centuries later, in 1979, that Pope John Paul II asked for a commission to reconsider The Galileo Affair. 13 years later, in 1992, almost 350 years after the condemnation, the Catholic Church, through the Pope, apologised for what happened in the Galileo Affair. On 31 October of that year, in the Vatican's adorned Great Hall, the Pontifical Academy of Science met with the

Pope. French Cardinal Paul Poupard gave the opening address, saying that the theologians who "attacked" Galileo did not understand that what is written in Sacred Scripture about the physical world is not literal, and that Galileo therefore suffered for the errors of his judgement. He also emphasised that the philosopher was more perceptive than the theologians of the time when it came to interpreting the Scriptures. Faced with these considerations, John Paul II accepted what the commission had said, recognising Galileo's suffering at the hands of the clerics and also the mistakes that had been made. However, at no point were the attitudes of Pope Urban VIII and the Inquisition directly criticised (RESTON JR, 1995).

According to Mariconda and Lancey (2001), until that moment it was as if Galileo's shadow was haunting the institution. This philosopher spent his life fighting for the autonomy of science; trying to make it independent from religion.

CHAPTER 4

THEORETICAL BACKGROUND

4.1 Theatrical text based on Ball's theatre

This work presents a Theatrical Text which was written with the precaution of presenting the format of a dramaturgy based on a professional play. To this end, we used the book Backwards and Forwards: A Guide to Reading Plays by David Ball (2011) as a reference. In view of the importance of David Ball's studies in the construction of this work, we then propose a discussion of the aspects that this author considers relevant in the elaboration of a Theatre Text.

Ball emphasises that at first the spectator has no knowledge of the play, so in the first scene important aspects should be explained so that the audience can build up a sense of place, historical temporality, characters and the social context in which the play is based. This will make it easier for the viewer to follow the story and the conflict in the play.

According to Bali (2011, p.17, emphasis added) "A (theatrical) text is not simply narrative prose in dialogue form. It is a modality of written language, which is extremely dependent on specific methods and techniques for the theatre."

In addition, the author considers that a Theatre Text must contain the idea of action. He emphasises that:

> **Action occurs when something happens that causes or allows something else to *happen*. Action is "two things happening", *one leading to the other*.** Something causes the action or allows something else to happen (BALL, 2011, p. 23 and 24, author's italics).

According to David Ball, one of the first things to discover in a Theatre Text is the course of the play. To do this, the first event of an action must be found, then the second event, which must be connected to the previous one. And then the third event, which must also be connected to the second. In this sense, the play will be assembled with its actions.

For Ball, the first event is called the Detonator, and this leads to another event called the Mount. The realisation of these events is an act of action. In this line of reasoning, the Mount will be the detonator of the next event, which will be the Mount of this one, thus providing yet another act of action. As a result, the Theatrical Text is made up of a series of actions, the actions in Bali's words being: "...the actions are like dominoes, each one falling on the next. The sequential analysis *8Í(Baica 8follows the play from START to finish, fromBinó to doBinó*"[8] (Ball, 2011, p. 29, italics added). In these sections, there is a sense of perception that is relevant to the proposal of arousing the viewer's attention. Because of this, it becomes a mechanism aimed at enabling the audience to pay the necessary attention to what will happen in the next scene.

Still according to Ball, action is the result of the actions the character takes to achieve what they want, as well as the obstacles inherent in their construction. Therefore, in order to get to know the character better, we need to find out what they want, the obstacles that exist in their development and what they do or are willing to do in order to achieve what they want.

The author also discusses issues related to stasis in theatre. For Ball, stasis is the moment of equilibrium between all existing forces, resulting in a lack of movement. In view of this, for the play to move, an Intrusion is needed, which consists of something that shakes the Stasis, causing or releasing forces, which act on the progression of the play and can provoke conflicts. Thus, the piece can begin with a Stasis and, during the piece, an Intrusion occurs, and at the end, the forces no longer collide, obtaining a new Stasis. Ball emphasises that:

> At some later point in the play, the most important forces of conflict must clash. This ultimate contest results in the restoration of an equilibrium - either the equilibrium that began the play, or a new equilibrium.
> ...as a gradual, step-by-step rise in intensity towards the shock, i.e. the **climax**, followed by a rapid fall in intensity during the final stasis (BALL, 2011, p. 127, author's italics).

Ball emphasises that there is one point on which everyone, or almost everyone, who talks or writes about theatre agrees: "Drama is conflict!". And dramatic conflict is

---

[8] The author himself wrote it in italics.

different from the others. The author exemplifies this argument by considering that:

> The conflict in a novel can be free will versus destiny. The conflict in a poem can be youth versus old age, or city versus country. But the conflict of a play is between what someone wants and what stands in the way of that want - the obstacle (BALL, 2011, p. 49).

According to Ball, there are four types of dramatic conflicts present in plays, which are entitled: Me against myself (1), Me against other individuals (2), Me against society (3) and Me against fate, or the universe, or natural forces, or God, or the gods (4). In this perspective of dramatic conflicts presented by Ball, the present work, which consists of the Theatrical Text on The Galileo Affair, was developed according to the last two dramatic conflicts mentioned above. In this sense, we consider the Theatrical Text on The Galileo Affair as an "I against society" due to the fact that Galileo had opinions contrary to what the Church defended in his historical context and the papal interference in the writing of the book *Dialogue.* We also consider it an "I against God" because it goes against the way people at the time spoke and believed in the Holy Scriptures, which *revealed* God's thoughts on the universe and movements.

4.2 Brecht's Didactic Play

Bertolt Brecht, the German playwright, in his book Studies on Theatre (1978) starts the discussion with the question: "Can today's world, despite everything, be reproduced by the theatre?"[9] (p. 5). He replies:

> As far as I'm concerned, this is precisely one of those questions that, when raised, immediately impose themselves. Long gone are the days when theatre was only required to reproduce the world in a way that could be experienced. Nowadays, in order for this reproduction to actually become an experience, it must be in tune with life (BRECHT, 1978, p. 5).

Still according to Brecht "...today's world can be reproduced, even in the theatre, but only if it is conceived as a world susceptible to change" (BRECHT, 1978, p. 7).

[9] Although Brecht begins the work with this question, it was formulated by Friedrich Durrenmatt in a lecture he gave, as Brecht himself states.

From this perspective, Brecht's work in the theatre sphere provided a new way of doing things by directing the language of theatre towards changing the society around him, which is why his proposal also focused on the intertwining of theatre and didactic action.

In view of the possibility of the theatre bringing about changes in the social structure, themes linked to everyday life and the political events of the time, such as wars, social struggles, religious issues and family issues, began to be discussed in the theatre. As a result of the incorporation of these themes, the actions of men began to be criticised, since these discussions led to discussions between men who also wanted to explain what was happening in their time in order to change it. In this way, people could philosophise and learn in an enjoyable way (BRECHT, 1978).

According to Oliveira (2007), the themes of didactic plays are intended to show social events in their dialectical process: entertaining and making people think. An active theatre has to be both science and art. From this perspective, the Didactic Play has the function of making its participants active and reflective, through the collective practice of art, which has the function of instructing in relation to moral and political issues. However, Brecht advocates not judging the idea of good and evil, since his proposal is aimed at stimulating the participants' critical and reflective sense so that they can make their own decisions autonomously.

Brecht's proposal is evidenced in a dialogue between a playwright and a philosopher. In this text, Brecht questions the idea of learning in the theatre through experiment alone. For him, it is necessary to discuss and reflect. At one point in the conversation, the playwright asks: "And how do you learn from experience? In theatre we don't just observe, we also experiment. Is there a better way of learning?" (BRECHT apud KOUDELA, 2010, p. 103). The philosopher replies:

> We don't learn anything from experience without the introduction of commentary. There are many moments that prevent us from learning by experimenting. For example, when certain changes in the situation happen too slowly, imperceptibly, as they say. Or when attention is distracted by other incidents. Or when the cause is sought in events that were not the cause. Or when the experiencer has strong prejudices.

> Playwright: Wouldn't he abandon his prejudices through certain experiences?
> Philosophers: Only if you've thought about it. Even then, you can still run into obstacles. Playwright: Isn't learning by doing the best school?
> Philosophers: In the theatre we don't just learn by doing. It is wrong to believe that every experience is an experiment and to want to take away from experience all the advantages that an experiment has. There is a huge difference between an experiment and an experience (BRECHT apud KOUDELA, 2010, p. 103).

We can see that, through the language of theatre, Bertolt Brecht used the word as an instrument of social transformation, since Brecht's Didactic Play was made for amateurs, not for theatre professionals, since most of his dramaturgies, following this model, were made with the intention of working with young people. For Brecht, there was potential in amateurs that they had to discover and find (KOUDELA, 2007).

In the following quote, Brecht justifies the importance of amateur theatre:

> We often forget how theatrical man's education is. The child experiences how to behave in a totally theatrical way, long before they are equipped with arguments. When this or that happens, you have to laugh. It laughs when it shouldn't and doesn't really know why. Most of the time, they are confused when you ask them why they laughed.
> He also cries with others. He doesn't just cry tears because adults do, but he also feels sincere sorrow when he cries. This can be seen at funerals, the meaning of which children don't grasp. These are theatrical processes that form character. People copy gestures, mimics and speech. And tears come from sorrow, but sorrow also comes from tears. Adults are no different. Their education never stops. Only the dead are not transformed by their equals. This explains the significance of theatrical play for character formation.
> Nor does art remain immune to the way it is exercised in its most common, carefree and naive manifestation. The art of theatre is the most human and widespread of all the arts, the one that is most practised, that is, the one that is not only exercised on stage but also in life. And the theatre art of a people or an era must be judged as a whole, like a living organism, which is not healthy unless it is healthy in all its members. This is also the reason why it is worth talking about amateur theatre (BRECHT apud KOUDELA, 2007, p. 20).

Another proposal is to use theatre in educational practice, since theatrical language is a playful way of learning (OLIVEIRA, 2007). According to Brecht:

> If it weren't for this possibility of a different kind of learning, theatre, despite all its structure, wouldn't be able to teach.
> Theatre never stops being theatre, even when it is didactic; and as long as it is good theatre, it entertains (BRECHT, 1978, p. 50).

In relation to the teaching and learning process of the Didactic Play, Koudela (1992) states:

> Brecht's theory of teaching and learning is a dialectical pedagogy that combines inductive and deductive elements in learning, providing us with a method of examining and acting on social reality. His concept of the **didactic play aims to teach the "joy of liberation" and make the act of** freedom **apprehensible** (KOUDELA, 1992, p. 12).

Steinweg (1992) warns that in practice there is no "one" method for working on the Didactic Piece. Just as there are no ready-made recipes for performing it. It is the objectives and specific needs of each situation that create the methods. However, there are common points in the Didactic Play that should be emphasised:

> ...the relationship with consciousness and the practice of everyday life, the dissolution of habits of perception, the work with social meanings that manifest themselves bodily, the play of role exchange as a means of identification, distancing and estrangement (STEINWEG, 1992, p. 48).

In Brecht's Didactic Play, the two main didactic instruments are: the "action model" and the "strangeness" effect. According to Brecht's proposal, the Theatrical Text is the "model of action", the mobile of action, the starting point for the imitations and criticisms that are introduced in the improvisations and discussions. At the same time as it is imitated, it is also criticised by the participants, and does not tell truths in itself.

"Action models" can and often must be isolated into parts in order to be judged and/or imitated. The "action model" proposes a relationship with some social case that is not necessarily associated with the participants' experiences. Some of Brecht's Didactic Plays, for example, are set in China, Rome, etc. However, the Theatre Text that follows the model of a Didactic Play does not build knowledge by itself, it provokes a process that can lead to knowledge (KOUDELA, 1992).

By understanding the "action models" of the Didactic Pieces as didactic instruments to be experimented with, we emphasise that they can be modified when new questions or new points of view are raised based on the situations in the text. These changes can be related to specific points or even expand issues beyond the text, thus bringing out new versions. In this way, according to Brecht, a new Didactic Play can be created by introducing new factors and enabling a "chain of experiments". Participants can modify the Theatre Text, but they have to try it out and discuss it first

(KOUDELA, 2007).

The Theatrical Texts of Brecht's Didactic Plays express social tasks in a dramaturgical structure (form) of their own. This form and politics are brought together, but not fused. Both can be perceived with their own particularities, but at the service of Brecht's aesthetic-political project, which manages to encompass and merge them because of the simultaneity between theory and practice (TEIXEIRA, 2013).

In practice, however, there is no method for working with the "action model". The needs of each experience generate their own procedures. The objectives of the activity, the age of the group and the specificities of the participants all influence the way in which the "action model" is approached. Participants will become active in the activity when they become active readers/actors. Thus, working with the "action model" becomes a means of improvisational theatre, a socio-aesthetic experience (KOUDELA, 2010).

The exchange of roles by the actors helps to avoid a single perspective. If the work with the "action model" is based only on identifying a single role, the Theatre Text will be interpreted with a one-sided view and there will be no perception of attitudes from different angles (KOUDELA, 2010). For Brecht, when the participant experiences acting with a "negative" behaviour, by identifying with the reverse, he builds knowledge about what he rejected, or was less familiar and distant from (KOUDELA, 2007).

Oliveira (2013) carried out a theatre activity with young people from a neighbourhood in Salvador- BA. During the development of the theatre entitled "Hip Brecht Hop", the young people read several of Brecht's plays, and at the end of the discussions, they imagined a different ending for all of them. Since their imaginations were related to their universes, their desires were to break with injustice through suggestions in the texts.

When you make a process or character strange, it means, initially, that you are removing from that process or character what is evident, known, manifest, to the point of provoking astonishment and curiosity about it. In addition, to make strange also means to historicise, managing to represent people and processes as historical, thus being transitory. It can also happen with more contemporary facts. Attitudes can also be characterised as temporal, historical and transitory (KOUDELA, 2007).

Brecht indicates some resources to provoke strangeness when reading the Theatre Text: transposing the speech to the third person or to the past and intrusions of indications about the staging and comments. By following these methods, the actor gains distance from the action, making it easier to form a concept and recognise that it can be changed (KOUDELA, 2007). In this respect, Brecht emphasises:

> The use of the third person and the past tense allow the actor to adopt a truly detached attitude. In addition, the actor should include in his performance indications about the staging and also expressions that comment on the text, uttering them **together with** the text **in rehearsal ("He got up** and said, grumpily, **because he hadn't eaten anything..." or** "He was hearing that for the first time, and he **didn't know if it was true...", or "He smiled and said, with too much nonchalance:..."). The intrusion of indications in the third person about how** to represent causes distance from the second person (the text itself). The representation will also be distanced if its actual realisation is preceded by a verbal description. In this case, the adoption of the past tense places the speaker on a plane that allows them to look back over their speeches. In this way, the speech is distanced without the speaker assuming an unrealistic perspective; in fact, unlike the audience, the speaker has already read the play through to the end and can therefore comment on any speech, starting from the outcome and consequences, better than the audience who know less than they do and who are therefore as if oblivious to the speech (BRECHT, 1978, p. 82).

With regard to distancing, Brecht (1967, p. 137) defined it like this: ""Distancing" a fact or character is, first of all, simply removing that fact or

all that is natural, known and evident about this character, and to give rise to awe and curiosity in its place."

Steinweg (1992) emphasises the significance of strangeness in the Didactic Play and the consequences of working well with this didactic tool:

> The meaning of strangeness is linked, in the didactic piece, to body memory. We don't just remember verbalised content; it's as if the body itself, torso and limbs, holds the social meaning of gestures, tones and body postures. Only when gestures and attitudes are perceived, touched and recognised in the context of their meaning and made repeatable (without losing the intensity of feeling) is it possible to behave differently in everyday life and in similar situations... (STEINWEG, 1992, p. 61).

The Didactic Play is made with the intention that the actors learn; not that the audience can't learn, but the main objective is that this happens to the participants.

Brecht argues that the Didactic Play should be based on the expectation that the actors will be socially influenced, so that they can reflect on their ways of acting, assuming (new) postures and reproducing (new) lines. And this happens when they perform certain actions, assume certain attitudes and repeat certain gestures. The group's artistic act happens when they imitate and criticise models of attitudes, behaviour and discourse. In the theory/practice relationship of the Didactic Play, there is a method of intervention of thought and action in the social issue. Thus, teaching and learning aims to generate critical attitudes and political behaviour (KOUDELA, 2007).

Brecht criticises the passivity created by bourgeois philosophers between the actor and the spectator, creating a great difference between them, and if this practice is maintained, the actor will be the politician and the spectator the philosopher. In this respect, Brecht emphasises that: "...when in reality politicians should be philosophers and philosophers, politicians. There is no difference between true philosophy and true politics." (BRECHT apud KOUDELA, 2007, p. 15). Recognising this, the participants have to be both actors and spectators of the Didactic Play (KOUDELA, 2007).

In the Didactic Play, the participant acts for themselves, not for others. At the moment he is acting, he is his own spectator. In this way, the Didactic Play teaches when you act and are a spectator of your actions. Because of this, the social influence of theatre happens when the participant assumes certain attitudes, performs certain actions and repeats certain lines. The idea of learning is the path to self-knowledge based on the principle of the activity of the participant, who becomes the artisan of their own education by managing to make the dialectical relationship between theory and practice (KOUDELA, 2007).

Steinweg uses the term "basic rule" to indicate the performance of the participant for himself as a prerequisite for the practice of the Didactic Play as an artistic act. From this idea, Steinweg understands the meaning of the statements made by Brecht, who argued that learning is achieved in the Didactic Play when the participant experiences it, not when he watches it (KOUDELA, 2007).

Although the Didactic Play doesn't need an audience, it can be shown. When he wrote Theatre Texts, following this model, Brecht thought of an attentive and nonpassive spectator in front of the art presented on stage. The spectator can't just feel emotions, he has to feel like an actor in the reality presented; having the ability to criticise and change certain situations (OLIVEIRA, 2013).

The cause and effect relationship between thought and its physical expression is the fundamental part of Brecht's concept of learning. Along these lines, in the Didactic Play he defends the gestus, which assumes significant importance in his aesthetics (Brechtian aesthetics). The idea of the gestus is intended to make what is subjective understandable and accessible, by clarifying the interobjective relationships that are constructed in social interaction. In the Didactic Play, the educational process aims, through the execution and observation of the participants' attitudes and gestures, to make them aware of their internal positions, externalising them (KOUDELA, 2007).

The gestus is intended to make what is subjective (subjective behaviour, subjective attitude) understandable from what is intersubjective, social. The social gestus relates attitudes and behaviour to reality. The models of behaviour formed by people are the result of a culture determined by their social class, gender, language, articulation, etc (KOUDELA, 2007). Brecht emphasises that:

> By **gestus** we mean a complex of gestures, mimicry and utterances, which are addressed by one or more people to one or more people.
> A man who sells fish shows, among other things, the **gestus** of selling. A man who writes his will, a woman who attracts a man, a policeman who beats up a man, a man who pays off ten men - all of these are part of the social **gestus**. A man invoking his God is only a **gestus**, in this definition, if he does so with a view to other men or in a context where there are relations between men and men.
> A **gestus** can only be expressed through words (on the radio), so a certain gesture and mimicry will be introduced into the words, which can be detected (a humble bow, a pat on the back) (BRECHT apud KOUDELA, 2007, p. 101).

In relation to the gestus, Brecht also emphasises that:

> ...when we talk about **gestus** we are not referring to gesticulation; we are not talking about hand movements in order to emphasise or explain speech, but about general attitudes. A language is gestural when it is based on **gestus**, when it reveals certain attitudes of the individual speaking, assumed towards other individuals... (BRECHT apud KOUDELA, 2007, p. 101).

The "educational effect" can be achieved through imitation. However, imitation is not restricted to the model in the text. Brecht emphasises that criticism of the text should lead to alternative ways of acting with inventions by the actors themselves, who are influenced by experiments outside the text, in the participants' reality. In this way, the text should only be updated when the participants relate it to their everyday experiences. Thus, the contradiction between

imitation and criticism dissolves when the model is allowed to be modified. In this case, the criticism may be contained in the imitation (KOUDELA, 2007).

Highlighting the discussions proposed in this text, we would like to highlight the ideas of Steinweg (1992), who considers that Brecht defended, based on the texts of the Didactic Plays, the experiences and ideas about the world and society of the participants in the theatre activity can be worked on and deepened in a way that only theatre allows. It only makes sense when gestures, body postures and movements related to space (all of which are specific objects of theatre) also take on some political and social significance in everyday life. Therefore, the Didactic Play aims to strengthen the individual as a social being, in order to prepare them for the transformation of situations and possibilities for action in their lives.

CHAPTER 5

METHODOLOGY

5.1 Nature of Research

This is a qualitative study. According to Minayo: "the qualitative approach cannot claim to reach the truth, with what is right or wrong; it must be primarily concerned with understanding the logic that permeates the practice that takes place in reality" (MINAYO, 1999, p. 18). Bardin (2011, p. 145) emphasises that: "Qualitative analysis has certain particular characteristics. It is valid, above all, for drawing specific inferences about an event or a precise inference variable, and not for general inferences".

5.2 Subjects

The research was carried out with 12 high school students from a public school in the state of Pernambuco, located in the municipality of Jaboatão dos Guararapes. The participants were selected voluntarily on the basis of a presentation of the proposed theatre activity. This school was chosen because the teacher/researcher was part of its teaching staff, making it viable and perfectly feasible to carry out the research at this institution.

5.3 Procedures

The research method had the following stages:

- Bibliographical review of Galileo's life and Brecht's Didactic Play;
- Construction of the theatre text "A Matuta e O *Caso Galileu",*

- ❖ Contacting the school management to obtain authorisation to carry out the activity;
- ❖ In the first part of the activity, lessons were given on Galileo's life, his contributions to science and his conflict with the Church;
- ❖ At the end of the last lesson, the students were given the Theatre Text, suggesting that they read it at home before the second lesson, following the idea of strangeness[10] from Brecht's *Didactic Play;*
- ❖ The second moment saw the presentation of how the debate would be conducted according to the ideas in Brecht's Didactic *Play.* The students/actors then read the text[11] , which served as a "model for action", sparking the debates. This moment was filmed;
- ❖ At the third stage, a questionnaire was administered to the research subjects ;[12]
- ❖ We analysed and interpreted the results.

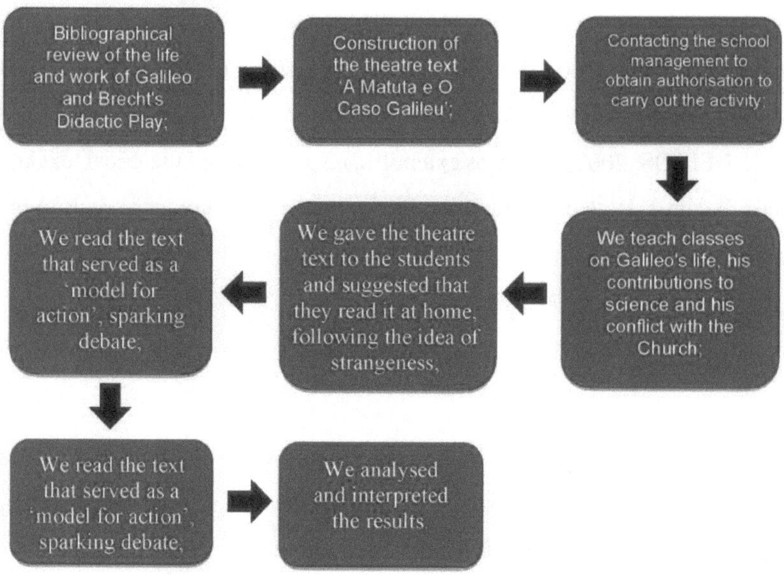

[10] The concept of estrangement is part of one of our theoretical references, Brecht's Didactic Play, according to Brecht's own readings and those of Koudela and Steinweg.
[11] Brecht refers to the actors in the Didactic Play as actors.
[12] The teacher/researcher did not stay where the questionnaire was administered. Another person, who was not involved in the activity, stayed in the room.

Flowchart 01 - Procedures for carrying out the research.
Source: prepared by the author.

5.3.1 The Theatre Text

The construction of the Theatrical Text (Appendix 01) was based on a study of the Italian physicist, mathematician and astronomer Galileo Galilei (1564-1642), which resulted in section *B* "A Brief History of Galileo Galilei with an Emphasis on the Historical Moment Known as The *Galileo Affair"*. This research formed the basis for the construction of the story created in the dramaturgy performed here, which emphasised the conflict that Galileo had with the Church and which became known as *The Galileo Affair*. The particularities of the Theatrical Text, which served as the appropriate construction for its elaboration, were based on the book *Para Trás e Para Frente: um Guia para Leitura de Peças Teatrais* (*Backwards and Forwards: A Guide to Reading Plays*) (BALL, 2011) and the idea of Brecht's Didactic Play, both of which constitute the theoretical basis for this work. This latter reference had a significant influence on the writing of certain moments of the Theatrical Text so that it could be constituted as an "action model". The book *The Life of Galileo* (BRECHT, 1977) and the article entitled *Interview with Kepler: from his Birth to the Discovery of the First Two Laws* (MEDEIROS, 2002) served as examples, respectively, of theatrical text and contextualised dialogue. Both deal with scientific knowledge and unique aspects of the life and work of scientists.

5.4 Research instrument

The main research tools in this study were the following: (i) the questionnaire that the students answered in the third moment; and, (ii) the transcript of the filming of the debates that took place when the Theatrical Text was read.

5.4.1 Questionnaire

The questionnaire that was applied, with eight (8) open questions, consisted of the research instrument that provided the analysed data.

Based on the answers given by the participating students, a set of conceptual categories emerged from the corpus obtained by applying the questionnaire and which best suited the corresponding comments.

With regard to categories, Bardin (2011, p. 148) emphasises that: "Classifying elements into categories imposes the investigation of what each of them has in common with others. What will allow them to be grouped together is the commonality between them". Flick (2013, p. 59) points out that "...identical or similar statements are summarised in a category. This leads to the development of a category system".

Finally, the answers were analysed.

With regard to analysis, Bardin (2011) points out that:

> The analyst is like an archaeologist. He works with **traces**... But traces are the manifestation of states, data and phenomena. There is something to discover through and thanks to them. Just as ethnography needs ethnology to interpret its detailed descriptions, the analyst takes advantage of the treatment of the messages he manipulates **to infer** (logically deduce) knowledge about the sender of the message or its environment, for example (BARDIN, 2011, p. 45, author's italics).

5.4.2 Transcript

The transcript of the filming was a research tool that, together with the questionnaire, contributed to the inferences. According to Bardin, inference is a: "logical operation by which a proposition is admitted by virtue of its connection with other propositions already accepted as true" (BARDIN, 2011, p. 45).

CHAPTER 6

ANALYSES

When the questionnaire was administered to the students/actors (6 men and 6 women), eight questions were asked. Based on the comments of the 12 students involved in the activity, the answers to each question were categorised in such a way that they were organised in a table with the identification of the students listed (1, 2, 3, 4, 5, 6, 7, 8, 9, 10, 11 and 12) so that the anonymity of each one was respected.

The aim of the questionnaire was to investigate whether, as a result of the activity carried out, the students had broken away, even partially, from some warped views of science, not seeing it as an absolute truth. Also, to see if the Theatrical Text can be considered a "model of action" so that the participants can debate/criticise the attitudes of the philosopher Galileo and the Church to the point of, at times, relating them to facts from the students' everyday lives.Right after tables 01, 04, 05 and 07, a transcript of part of the debate held during the activity was added, in order to contribute to the inferences made from the categorisation data. In addition to these four excerpts from the discussion, at the end of this section, another transcript has been added because in this specific debate some warped views of science that were not mentioned in the analysis of the eight tables are broken down, even partially, at this point.Below are the tables and their respective analyses for the purpose of contextualising the approaches inherent in the didactic and pedagogical process in the Teaching and Learning of Physics Teaching in a 3rd year secondary school class.

TABLE 01 - Categorisation of answers to question 01:

In order to publish his book Dialogue on the Two Maximum Systems of the Ptolemaic and Copernican Worlds, **Galileo went to Rome to take care of the formalities of publishing this work. In Rome, Father Riccardi made the first revision and Father Raphael Visconti made the corresponding revision of the scientific part.** Castelli then suggested that Galileo travel to Florence to speed up the process of publishing his book. Why did Castelli and Galileo prefer to publish the book in Florence rather than Rome?

Categories of Answers	Students	Frequency of responses	%
In Rome, religious pressure was much more intense than in Florence. Galileo was right to accept Castelli's advice, because if the book was published in Rome, it would be easier for the Church to	1, 2, 4, 5, 7, 8, 9, 11, 12	9	75

intervene.			
In Rome, with the intervention of the Church, the process of publishing the book would take longer.	3, 6, 10	3	25

At this time in the history of science, the Church was the most powerful institution in society and defended the Aristotelian-Ptolemaic theory, not accepting other ideas as "truths". For this reason, it wasn't going to be easy for Galileo to publish his book defending Copernican ideas, a theory that went against that defended by the Church, especially in Rome, which was, and still is, the city where the Vatican is located. This shows the relationship between this scientific knowledge and society, breaking the decontextualised view, according to Cachapuz et al. (2005), which is when science and technology are talked about without relating them to society. Looking at the students' answers, they reinforce this break when they recall that it would have been easier and quicker for Galileo to publish his book in Florence because in Rome the Church could have intervened more significantly in this process. After reading the first scene, there was a debate between the students and the teacher, reinforcing the idea mentioned above.

Student 11: He (Galileo) knew that the Church was going to stick its finger in.

Student 02: Faced with the conditions imposed...

Student 09: Has he talked to the Pope about this?

Professor: Already. At that point he had already spoken to the Pope. The Pope had already authorised Galileo to write a book on the two systems hypothetically.

Student 03: In that part he's going to publish the book in Florence to... since he can't... he's here (Rome)...

Professor: He's in Rome at the moment. The text was analysed by Father Riccardi and the scientific part by Rafael Visconti. After the analysis, Castelli suggested that he (Galileo) go to Florence and publish the book there.

Professor: What did you think of this tactic of leaving Rome and publishing the book in Florence?

Several people speak at the same time.

Student 03: It was a joke.

Student 01: A manoeuvre.

Student 08: It was good.

Student 06: It was a necessary evil.

Student 03: He kind of... jerked the Church around. Because... he was waiting for an order to publish the book. So he went somewhere else and published it there. And like... fuck the Church. **Student 08**: Because... if he published it there (in Rome) he'd get screwed. **Professor**: Would it be right for him to stay in Rome and publish there? **Student 03**: Yes. **Student 10**: Now... the two priests who read his text... didn't comment on anything negative. **Student 01**: But it was the beginning and the end. It was the beginning and the end of the book, right? **Teacher**: When he went to Florence, the Church asked him once again to analyse the text. At that time, it only analysed the beginning and the end.

In this dialogue, the students reinforce the idea that the Church interfered in this historical moment in science. Breaking, as has already been said, the deformed view of the Decontextualised form. In this debate, we can see the students analysing this part better. A better understanding of their reasoning. By saying that Galileo knew that the Church would put its finger on it, and remembering that the Pope had authorised the Florentine to write the book in a hypothetical way, shows that the students realised that Galileo made the decision to publish in Florence because he knew that there was a great chance, if it were in Rome, that the Church would interfere in the publication
of his work. Most of the students defended Galileo by saying that he had to do it. He had no other choice. However, student 03 criticised the philosopher, saying that he had screwed over the Church, and that he had to publish the book in Rome, not go somewhere else to publish it. This last student's thought is important for looking at the Theatrical Text as a "model of action", since it led the students to debate Galileo's attitude, as well as criticise him in a way that some agreed with and others didn't. Thus, they began to see Galileo as a philosopher. In this way, they began to see Galileo as a normal person who gets things right and wrong. The "action model" is a didactic tool in Brecht's Didactic Play that aims to provoke the actors to debate the attitudes of the characters in the Theatre Text.

TABLE 02 - Categorisation of answers to question 02:

In his book Dialogue on the Two Maximum Systems of the Ptolemaic and Copernican Worlds, **Galileo created three characters who discuss various pieces of knowledge related to physics and astronomy. Briefly, describe the characteristics of each of these characters.**

Categories of Answers	Students	Frequency of responses	%
The character Simplicio, named in a derogatory way, defended the ideas of Aristotle and Ptolemy and was often challenged by the character Salviati (who represented Galileo's heliocentric vision). The third character always agreed with Salviati, leaving the argument two against one.	1, 2, 3, 4, 5, 6, 7, 8, 10, 11	10	83,33
One was Galileo or his conscience. Another followed more of the Church's theories. The third was more of a layman.	9	1	8,33
Simplicio was considered by many to be representing the Pope and to be a form of **"debauchery". Another character went against** Simplicio's ideas and the third was a **"middle ground" between the two.**	12	1	8,33

When writing his book Dialogue, Galileo created three characters who discussed the theories of geocentrism and heliocentrism. However, in this debate, the author showed, with arguments, that heliocentrism was the most coherent theory with the discoveries and knowledge of the time and revealed the weaknesses of geocentrism. Thus, according to Cachapuz and collaborators (2005), we can see that the deformed view of science known as Accumulative is at least partially broken. According to these authors, this distorted image of science occurs when scientific knowledge is presented as linear growth. The "fights" between rival theories *and the* problems encountered are not commented on. The students reinforce this "break" by talking about the characteristics of the three characters created by Galileo in his book, saying that Salviati defended the author's ideas and heliocentrism, Simplicio defended geocentrism and Sagredo, in the end, always agreed with Salviati, reinforcing the defence of the heliocentric theory. In this way, at that moment in history there was a "heated dispute" between these theories, due to the new ideas that were contributing to the break-up of the scientific knowledge defended at the time, in this case the Aristotelian-Ptolemaic theory, and thus the strengthening of the study of a new theory, in this case heliocentrism, which was in line with the new knowledge.

TABLE 03 - Categorisation of answers to question 03:

When Pope Urban VIII authorised Galileo to write a book talking about the two main theories of the universe, the Ptolemaic (geocentric) and the Copernican (heliocentric), in a hypothetical way, did Galileo follow what was

requested by this Pontiff? (justify your answer)

Categories of Answers	Students	Frequency of responses	%
He only followed the title of the book; Galileo didn't follow the hypothetical form to the letter, which led to his condemnation.	1,2	2	16,67
No. He didn't present the theories hypothetically in his book. Through a character, he defended the Copernican theory as true and real.	3, 4, 5, 6, 8, 10, 11, 12	8	66,67
For me, yes, because I think **Galileo**'s point **about "hypothetical" was the creation of a** plot between characters.	7, 9	2	16,67

In order for Galileo to write his book on Copernican theory, he needed authorisation from the Church. Therefore, this moment in history shows the relationship between science and the Church, which was the institution with the greatest power in society and which had a great influence on people's lives. As a result, it can be said that there was a break with the deformed view of science, known as Decontextualised Science, according to Cachapuz et al (2005). The students' answers reinforce this statement when they confirm the Pope's interference in the writing of Galileo's book. Thus, the Pope's attitude in condemning Galileo was fundamentally due to the fact that the famous Florentine had disobeyed his advice to present the geocentric and heliocentric theories in a hypothetical way.

At that moment in history, the heliocentric and geocentric theories were "fighting" to show which one best explained the movement of the stars. Therefore, another deformed view of science that is at least partially broken here is the Accumulative view, according to Cachapuz and collaborators (2005). The students' answers confirm this idea by showing that Galileo wrote a book talking about the two theories, but defending heliocentrism. For this reason, he ended up contributing to his own condemnation, since the Church defended the ideas of Aristotle and Ptolemy and, as mentioned above, the Pope authorised this philosopher to write a book talking about the two theories in a hypothetical, non-biased way.

TABLE 04 - Categorisation of answers to question 04:

After publishing the book Dialogue on the Two Maximum Systems of the Ptolemaic and Copernican Worlds, **Galileo was summoned by the Inquisition. What were the reasons that led the Pope to make this decision regarding this** philosopher?

Categories of Answers	Students	Frequency of responses	%

Galileo didn't fulfil the Pope's request and they put it into the Pontiff's head that the character Simplicio was inspired by him, which made Urban consider it an affront to religious power.	1	1	8,33
The Pope turned the summons to Galileo into a personal motive.			
Due to pressure from some within the Church, the Pope was led to believe that the character of weak arguments and questionable knowledge was based on the Pontiff. During this period, the Pope was being criticised for not being tough enough on heretics. All this led the Pope to become angrier with Galileo, making his situation increasingly difficult.	2	1	8,33
0 reason was that Galileo didn't fulfil what he and the Pope had agreed, which was to write about the two theories hypothetically. On the contrary, Galileo defended the Copernican theory and ridiculed the Aristotelians. Another reason that led him to the Inquisition was that the Pope felt humiliated because his order was not obeyed and he was told that the character Simplicio was the Pope himself, a character who was seen in such a way that his ideas were ridiculed and belittled.	3, 4, 5, 8, 9, 11, 12	7	58,33
The way in which Galileo approached the theories set out in the book was different from the way in which the Pope had allowed him to approach them (i.e. in a hypothetical way); hence, summoning him to the Inquisition.	6, 7, 10	3	25

After publishing the book Dialogue, Galileo had problems with the Church to the point where they banned the sale of his work and summoned him to the Inquisition. There were several reasons why the Pope made these decisions; showing the interference of this institution in the research and publications made by the philosopher Galileo. Because of this, this moment serves to break the deformed view of Decontextualised Science, according to Cachapuz and collaborators (2005). The table above reinforces this idea with the students' answers when they affirm this attitude taken by the Pope, completing the reasons with the corresponding whys. With these motivations, they consider Galileo's failure to comply with the Pope's request to have been of decisive importance. All the more so when all this was added to the Jesuit intrigue (Scheiner and Grassi) of putting it into Urban VIII's head that the character Simplicio, who had foolish ideas, was exactly based on real speeches by the Pontiff himself. In addition, the students claim that the Pope was being criticised for not being tough enough with heretics, and that the Galileo affair could therefore be an excellent opportunity for him, as Pontiff, to rigorously defend the Christian faith.

In the reading of the Theatre Text, right after scene IV, there was a debate that reinforced the students' ideas above. See the extract below:

Student 01: It was a joke.

Teacher: What would be the fun in that?

Student 01: The cardinal kept putting things into the Pope's head.

Students 02 and 06: Yes!

Professor: Actually, the cardinal is a character I created. But... those things he said influenced the Pope. So, what were these things that influenced the Pope?

Student 01: The character.

Student 08: The character Simplício.

Student 09 nods in agreement.

Student 02: And his way of writing, which went beyond the Pope's order.

Student 06: In those days, a good priest was a bad priest.

Some students start laughing.

Teacher: What were you saying about Simplício?

Student 08: Like... he (the Cardinal) told the Pope that the character Simplicio was the Pope himself. That the Pope was a fool, silly,...

Student 03: He (the Pope) was the symbol of the Catholic Church.

Professor: This idea of Simplicio being the Pope himself, here in the Text it was the cardinal who said it, but some people tried to put it into the Pope's head. It even mentions the names of two people: Scheiner and Grassi. Two people who tried to do this and that Galileo had created this character with the intention of ridiculing the Pope.

Student 01: One of these is the one about...

Student 02: Sunspots.

Student 01: Sunspots. Scheiner?

Professor: Yes.

More on this later in the debate:

Student 03: It's like the cardinal said. People used to say that he (the Pope) was always "nice" to heretics. He couldn't be like that. He was the greatest representative of Catholicism. He couldn't disappoint.

Professor: In this situation, what would the Galileo case be for him?

Student 01: Everyone was on him because he...

Student 02: Because it messed with him, with his authority. He asked him to do one thing and Galileo did another. And he even ridiculed him.

Professor: He (the Pope) had to do something. In this case, he had to show the others... **Student 02**: I had to show authority.

This dialogue reinforces what has already been said above in relation to the students' answers in table 04; helping to break down the deformed view of Decontextualised Science. However, the students recalled an important detail in order to understand why Father Scheiner was one of those who tried to put it into Urban VIII's head that the character Simplicio was the Pope himself. The priest was an enemy of Galileo from the moment Scheiner accused the philosopher of saying that he had seen the sunspots before him. According to the priest, he had already informed Galileo that he had made this observation, and yet Galileo published this discovery as being the first he had seen. Because of this disagreement between them, they became enemies to the point that Scheiner, after the publication of the book Diálogo, tried to convince the Pope that the character Simplicio had been inspired by the Pontiff, contributing to the discomfort between Urban VIII and Galileo. Another detail is that the students also emphasise that the Pope was the greatest representative of the Catholic Church, and therefore he had to show authority, since Galileo had overridden his order.

TABLE 05 - Categorisation of answers to question 05:

At one point in the play, Matuta is outraged by the authority of the Church and vents: Matuta: **Why do these Church people have the authority to say what is right and wrong? Why have they acquired so much power that they can forbid a person to publish their ideas? Why can they torture and burn people alive? I consider all of this to be absurd! Do you know what I'm going to do?**

Galileo: **What?**

Matuta: **I'll pray for you.**

What do you think of Matuta's questions and attitude?

Categories of Answers	Students	Frequency of responses	%
The matuta questions the authority of the Church at a time when a large part of the population was subject to it. These are opportune questions that later began to be asked by other people. However, the character contradicts herself, as she criticises the Church but tells Galileo **that she will "pray" for him.**	1, 11, 12	3	25
She was a person of faith who didn't support the Church's attitude. It's surprising, because people at that time were usually alienated and thus subjected to the Church.	2	1	8,33

The matuta as a character who, like the spectators, is outside the Galileo affair, asks herself the same questions as the spectators, why so much power and so much imposition, questions whose meanings don't matter so much since they were so used to the Church with all its power in leadership. That's why the act of praying, which we've been taught since we were young, came naturally to her.	3	1	8,33
At that time, the power was with the Church, which decided everything, so as Galileo's book contradicted the Aristotelian-Ptolemaic idea, defended by the Church, coupled with misinterpreted biblical accounts, it was impossible for Galileo to convince the Church that his theory was right, even with all the evidence available, and as it happened being condemned for heresy was to be expected, and no one could do anything to change that reality, not even the Matuta who believed in him.	4	1	8,33
In fact, at that time the Church was overstepping its bounds, taking attitudes that even went against the Holy Scriptures; the attitudes taken by the Church were absurd. As for the attitude taken by the matuta, perhaps it was an act of faith, not in the Church, but in God.	5, 6, 7, 8, 9, 10	6	50

This part of the play served as an example of a "model of action", a didactic tool in Brecht's Didactic Play, as it led the actors to defend and question Matuta's attitude, while also relating the situation to their everyday lives. Some students criticised the character for praying after questioning the Church, saying that she was contradicting herself. Other participants understood these two acts differently, saying that Matuta had questioned the Church, but that her attitude of praying was not related to that institution, but to God, and therefore she was not contradicting herself.

Matuta's attitude of praying for Galileo is called a gestus in the Didactic Play, since she prayed not for herself, but for someone else.

Some students were careful to point out that at that time it wasn't normal for people to question the Church so much. Others also said that at that time the Church had the power. It can be seen in these answers that the participants analysed Matuta in the Renaissance period, and not now.

Some students also took the opportunity to question the attitudes that were taken by the Church at the time, saying that it crossed the line and that the institution itself went against what is written in the Holy Scriptures. Student 3 reveals that the questions asked by the character Matuta are the same ones that many viewers would ask: Why so much power? Why so much imposition? It's as if the character Matuta was speaking for the people.

Here's when the students began to discuss this part of the Theatre Text after reading Scene V.

After finishing scene V, some of the students start laughing. **Teacher**: Why did you laugh? **Student 08**: Because Matuta spoke badly and then went to pray. *Several people speak at the same time.* **Student 03**: I want to know the answers. **Teacher**: What answers? **Student 03**: From her questions (Matuta) *Some begin to laugh.* **Teacher**: Let's repeat that last part with the questions: *Why do these Church people have the authority to say what is right and wrong? Why have they acquired so much power that they can forbid a person to publish their ideas? Why can they torture and burn people alive? I consider all of this to be absurd! Do you know what I'm going to do? You know what I'm going to do? I'm going to pray for you.* **Student 06**: Like... the Church used to do that... she goes to talk to the "boss". **Student 09**: If it exists, right? **Student 06**: But back then God was the "boss"... **Student 10**: It still is. For many. **Student 06**: That's what they said. **Student 10**: It still is. For many. **Student 08**: For many. **Teacher**: OK, but I still want to know, why did you laugh? **Student 08**: She talks about the Church and then goes to pray. **Student 01**: The woman screws around in church and then goes to pray. **Student 02**: She blasts the Church and then goes to pray. **Student 10**: Because her prayer would certainly be very productive. *Some begin to laugh.* **Teacher**: Do you think she said one thing and acted in a completely different way? **Students 01, 02 and 08**: Logical. **Teacher**: But can't she pray and at the same time not agree with the Church? **Student 02**: You can. **Student 01**: You can. **Student 03**: Yes, religion and... **Student 10**: It's because it looked like she was going to criticise and do something. But she went home to pray. **Professor**: Does this still happen today?

Student 02: Logical. **Student 03**: Logical. **Teacher**: Example? **Student 06**: You have cancer... I'm going to pray for you. *Some begin to laugh.* **Student 09**: Or else... like... you come here... I'm going to pray for you... I'm going to save you. **Student 01 and 08**: Yes! That's right.

At the start of the dialogue, the students analysed that at that time the vast majority of people believed in God and he was a kind of boss in their lives, and that many people still have the same thought today. In this way, they were able to relate the situation in the theatre text to today's reality, to the point of giving examples of situations of people who believe that through prayer some problems can be solved through the interference of something divine. In this debate, the students once again questioned Matuta's attitude of criticising the Church and then going to pray. For many, it was a hypocritical attitude. However, some accepted the idea that she could go and pray even if she didn't agree with the Church's attitudes.

These details reinforce the idea of looking at this Theatrical Text as a "model for action", to the point where the actors not only criticise attitudes, but also relate situations from the play to today's reality.

TABLE 06 - Categorisation of answers to question 06:

Regarding the attitudes of the Church and Pope Urban VIII in the Galileo Affair, in your opinion, did they err in this process? If so, what was the mistake(s)? If not, why?

Categories of Answers	Students	Frequency of responses	%
They were wrong to turn what should have been scientific into a personal affair. Let it be said that Galileo didn't fulfil what was asked of him and the Church was right to pass judgement. However, they turned Galileo into a scapegoat.	1,2,6	3	25
Yes, because he (Galileo) had evidence and arguments to prove what he was saying, even contradicting the misinterpreted biblical accounts, as he himself said. The Church judged and condemned him, mainly at the insistence of the Pope.	3, 4, 5, 7, 8, 12, 10	7	58,33
It was all a conspiracy against Galileo, since they deceived the Pope by claiming that a character in the book, with foolish ideas, was based on the Pontiff.	9	1	8,33
Yes, but there is a whole context around this. They (the churches) are institutions that move and change the minds of multitudes. They couldn't simply change from water to wine a theory they'd been	11	1	8,33

defending for years. It would cause people to doubt and they only work with absolute truths.			

When the Church decided to summon Galileo to the Inquisition and banned the sale of his book, it interfered in the development of science at the time, which was in the process of breaking down one theory and discovering another. However, with the intrusion of this religious institution, we realised the relationship that existed between science and society. Thus, breaking the deformed view of Decontextualised Science, according to Cachapuz and collaborators (2005). The students reinforce this idea by saying that the Church did interfere; showing and judging the attitude taken by it and Galileo in the context of the time. Showing that this issue makes the Theatrical Text an "Action Model", a didactic instrument of Brecht's Didactic Play, because the students can criticise the attitude taken by the Church towards Galileo, as well as his own. This can be seen when the students say that the Church made a mistake, especially Pope Urban VIII, in turning The Galileo Affair into a personal case, and also that it turned Galileo into a scapegoat to show the Church's critics that the Pontiff was tough on heretics. She also made a mistake because Galileo had evidence and arguments to prove many of the things he said. That's why the Church shouldn't have been so rigid in this case. But, as student 11 said, the Church couldn't change a theory from one moment to the next. That could cause doubts in its faithful. And the Church works with absolute truths. But some students pointed out that Galileo didn't obey an order from the Pope, who is the highest authority in the Catholic Church. Thus, he gave reason to be judged.

TABLE 07 - Categorisation of answers to question 07:

When Galileo was summoned to the Inquisition, he sought out some authorities to help him through the process. In the Theatrical Text there is a simulation of a conversation between him and Niccolini during this period. The latter makes a suggestion and ends up being surprised by what Galileo says in the dialogue. Watch this passage below:

Niccolini: **You could talk to the Venetian government to help you. Didn't you show them the telescope you invented?**
Galileo: **Invented by whom?**
Niccolini: **For you, wasn't it?**
Galileo: **Not for me.**
Niccolini: **But you told them that.** Galileo: **Yes.**
Niccolini: **And why did you say that?**
Galileo: **What I wanted was money! I couldn't research with all the private lessons I had to take to pay my bills. I had to pay for my two sisters' weddings, I helped my mum and brother who loved to spend and I had to support**

my three children.
Niccolini: **Galileo, you surprise me every day.**
Galileo: **Rubbish. It's nice to have that kind of winding up.**

What do you think of Galileo's attitude?

Categories of Answers	Students	Frequency of responses	%
It was the wrong thing to do, but he had the right reasons for doing it. I don't consider Galileo to be a bad person, but rather a human being who, like everyone else, wanted to improve his life. He preferred the illicit path to the licit one.	1,6,9	3	25
Ridiculous, a man of such greatness shouldn't put himself through this.	2,5,7	3	25
In a way, it was clever and smart, but at the same time it wasn't fair, because it wasn't his invention, so it was a "low blow", and he did it because he needed the money.	3,4, 12	3	25
The situation in which Galileo found himself led to this decision, as it was more advantageous for him to receive this information through the telescope than through his private lessons.	8, 11	2	16,67
He was clever, he took the initial project, improved it and showed it to the authorities.	10	1	8,33

When Galileo built his telescope, it wasn't his invention, as this instrument already existed and he built a better one than the one he had. However, as the Venetian government didn't yet know about this instrument, he took the opportunity to show them his and tell them what advantages the military would have with the telescope. Galileo thus took advantage of an invention that wasn't his own, doubling his salary and earning a bonus. This episode is an example of getting to know the person Galileo was, and seeing him as an ordinary human being who, at the same time as gaining an advantage in this episode, helped his mother, his brother and paid for his sisters' dowries. The students confirm this idea by saying that even though they didn't agree with his attitude, he had his reasons and was in need of money. They also say that he was an ordinary human being who wanted to improve his life.

Students 2, 5 and 7 responded more harshly to Galileo's attitude, saying that a man of his greatness shouldn't take such an attitude. This detail of the answer ("a man like him") leads us to believe that Galileo was not seen by these students as an "ordinary" human being, but rather as someone with a prominent position in history who could not take certain actions like "ordinary" men. Despite being the minority of participants, their view of men like Galileo seems to lead to the idea that scientists like him are not ordinary human beings who get things right and make mistakes. However, this example contributed to seeing this part of the Theatre Text as a "model of action", a didactic instrument of Brecht's Didactic Play, since it encouraged the actors to debate the scientist's attitude by observing his mistakes and successes.

After reading Scene IX, which contains the passage in Question 07 from the dialogue between Galileo and Niccolini, the students, together with the teacher, debated the scientist's attitude.

Professor: Well... there's a part here... has anyone ever imagined a famous scientist doing what Galileo did?

Student 02: Yes, it's a smart world.

Student 01: Yes.

Student 03: He used someone else's discovery to go up. He didn't invent the telescope.

Teacher: Can you imagine Socrates, Plato, Descartes, Newton walking all over each other?

Student 01: Yes.

Student 02: It's a smart world.

Student 01: If you have a greater interest, you'll pass.

Student 09 shakes his head in disagreement.

Teacher: Can you imagine Galileo doing that?

Student 02: It's not natural, but...

Student 10: It was necessary.

Student 08: I think it was necessary.

Teacher: Why?

Student 03: He needed money.

Student 02: Yes, but it doesn't justify it.

Student 02: The person is hungry, so they go and steal. It's not justifiable.

Student 03: You're at a job interview, you're hungry, aren't you going to try your best to get that job? You're starving.

Student 06: I think that regardless of how rational human beings are, there's the survival instinct that can do anything to survive.

Student 03: Exactly. He supported the family. He paid dowries, he had to pay...

Student 02: No justification.

Teacher: Because of his needs, did he really have to do that?

Student 03: There was.

Student 01: Out of necessity. **Student 08**: Out of necessity. *Student 02 shakes his head in disagreement.* **Student 02**: I only know that I know nothing. *Some students start laughing.* **Teacher**: would you? **Student 08**: I would. **Student 01**: It depends on the situation. **Student 02**: I don't know. **Professor**: Galileo was no pauper. **Student 03**: He was naughty. *Some begin to laugh.* **Student 01**: He was clever. He saw an opportunity and acted.

Initially, most of the students confirmed their thoughts in relation to seeing Galileo's attitude as something relatively normal for a human being. Then they justified it by saying that it was necessary for him to do this because he needed money; they even remembered his family that he helped and the dowries he had to pay. However, student 02 disagreed, saying that this was not a justification for this kind of behaviour. Student 03 tried to justify Galileo's attitude by saying that a person who is hungry ends up out-competing someone else for a job. Student 02, on the other hand, doesn't accept the justification. This discussion reinforces the idea of seeing the Theatrical Text as a "model for action", since it encouraged the students to criticise Galileo's attitude. And when the teacher asked them if they would "do" something similar, some of the students said what their attitudes would be.

TABLE 08 - Categorisation of answers to question 08:

One of the reasons for the conflict between Galileo and the Church was that the Church believed in the geocentric theory of Aristotle and Ptolemy, determining this knowledge as the truth, and Galileo defended the heliocentric theory of Copernicus. At this point in history, two ways of explaining the universe were being discussed. Based on this, do you think that scientific knowledge is an absolute truth? Why?

Categories of Answers	Students	Frequency of responses	%
No. There is always a new theory to overturn a previous one, which further reinforces that science is not an absolute truth.	1,3,7	3	25
No, scientific knowledge is not an absolute truth, because just as Galileo proved that the ideas of Aristotle and Ptolemy were wrong, other scientists can return to this discussion and prove Galileo	2,5,8, 11	4	33,33

wrong. Everything is constantly changing, and I believe the same is true of ideas.			
Yes, because the idea of geocentrism defended by Aristotle and Ptolemy was based on everyday facts without in-depth study and proof. Galileo and Copernicus, on the other hand, had facts and arguments to prove their theory. Scientific knowledge is studied and	4	1	8,33
proven, so it's an absolute truth.			
Truth is subjective for each being. The blindfold that covers the eyes of science is different from the blindfold that covers the eyes of the religious. Human beings in their current mental and psychic state do not have the capacity to see what many consider to be non-existent; each being has their own truth that often cannot be proven.	6	1	8,33
Yes, because most of the time what is questioned is proven.	9, 12	2	16,67
No. Much is still vague in terms of scientific knowledge. According to research, we know about 5 per cent of the universe, if it isn't infinite, so we can't make **scientific "truth"** absolute.	10	1	8,33

At this point in history, the geocentric theory, which was seen as "the truth", began to be questioned more intensely. The heliocentric theory began to gain more strength. Soon, what had been seen as "the truth" began to be replaced by a new theory. This historical moment serves as an example of the fact that scientific knowledge is not absolute truth, since new theories can appear that explain things in a better way than the previous one. In view of this, there is a break from the deformed view of science entitled Aproblematic and Ahistorical, according to Cachapuz and collaborators (2005), which is when science is taught as "ready-made" knowledge; being an absolute truth, not showing the development of scientific knowledge and its history. Besides not showing that "all knowledge is the answer to a question" (BACHERLARD apud CACHAPUZ ET AL, 2005, p. 49). The students' answers reinforce this "break" when they say that there is always a new theory that can overturn a previous one, thus showing that science does not have absolute truths. They complete this idea when they use the story of the Theatrical Text itself to say that just as Galileo helped to overthrow the ideas of Aristotle and Ptolemy, other people can show that those of the Florentine philosopher are wrong. Two other comments also reinforce the view that scientific knowledge is not an absolute truth. However, they respond in a different way. Student 10 sticks to the idea that human beings know only a small piece of the entire universe, and therefore cannot claim to say what is really true. That humans still have a lot to unravel in order to really know the whole universe. Student 6 is attached to the idea that there is not only scientific knowledge, but other knowledge too. And that each human being has their own "truths" that can be related to scientific knowledge, as well as religious or other types of knowledge.

Once again, according to Cachapuz and collaborators (2005), the deformed view of Accumulative Science is broken by once again showing the "fight" between the heliocentric and geocentric theories. By replying that Galileo contributed to overthrowing the ideas of Aristotle and Ptolemy and by saying that there is always a new theory to "break" a previous one, they reinforce the idea that there are moments in the History of Science when theories "fight" over which of them explains "better" what is being studied.

In the debate with the students, which was filmed and transcribed, there was a moment, in addition to the others that have already been discussed, that complements the analysis of this research by breaking some views that have not yet been mentioned. This dialogue took place right after scene III.

Professor: Here I have two arguments against geocentrism. One is the question of the sphere of the stars rotating in 24 hours or the Earth rotating around its axis in 24 hours. What does he say here that complicates the idea of the sphere of the stars rotating in 24 hours?

Student 01: The whole universe is spinning in 24 hours.

Student 10: The speed is very high.

Teacher: We have two possibilities here: either the whole universe rotates in 24 hours or the Earth rotates in 24 hours. Which would be more logical?

Several students answer: The Earth.

Professor: The Earth would be simpler. That's one of Copernicus' arguments. The other (argument) is the idea... the imaginary experiment inside a ship. How does this imaginary experiment relate to us on Earth?

Student 06: The Earth is the ship...

Student 08: We're on the ship.

Teacher: We're standing on the Earth. Do we realise how it moves?

Student 02: No.

Teacher: Here on Earth, if I throw something up, it will come back and I'll catch it. Disregarding air resistance. But everything is spinning (the Earth and what's on it), everything is moving. Now let's think about this idea on the ship. The ship is moving. Inside a compartment there are the birds, the little fish in the aquarium, the drop of water... what relationship did he want to make between the ship and the Earth in this experiment?
Student 09: I think it's the ship that's moving and the Earth that's spinning. You can't tell.
Student 02: The ship is moving, the Earth is standing still, the things on the ship are standing still, we're standing still.
Teacher: But aren't the birds flying inside?
Student 02. Yes, but they're with the ship.
Student 10: Like... on Earth... in relation to the Earth I'm at rest. In the case of the ship too. Just like on Earth.
Teacher: If we look at these birds when the ship is at rest and when it is travelling in a sea without jolting at a constant speed, can we see any difference?
Student 10: Not on the ship.
Student 02: From the outside you can see.
Student 02: On the ship, I don't think so because we're moving with it. It's in the same situation.
Student 10: The question is the referential.
Professor: You made an important point. We're in the same place... ship. When we're inside the ship, everything inside it is moving with it. And everything on Earth is moving with it. Galileo made this comparison. What was one of the arguments of the Aristotelians? If we threw an object upwards, it would move with the Earth and the object would not return to our hands. Since this didn't happen, then the Earth wasn't moving. It was an argument to say that the Earth didn't move. So Galileo thought: here's what we're going to do: imagine this situation with the ship that's here. Then, if it doesn't happen on the ship, take the same situation on Earth. He did this imaginary experiment to contradict the Aristotelian argument.

In this dialogue between the students and the teacher, Galileo's imaginary experiment is discussed, so that the students understand the relationship between the ship and the Earth when they realise that the philosopher's idea shows that in the same way that no difference can be seen in the movement of animals or the drop of *water* in the bucket inside the compartment of a ship moving in uniform motion, the same thing happens on Earth even though it rotates around itself at a relatively high speed. Thus, in this debate there is a break with the deformed vision of Individualist and

Elitist Science, according to Cachapuz and collaborators (2005) who state that this type of image of Science is when it is shown as the knowledge of a few. It is isolated geniuses who manage to understand it. Just as the discoveries of scientific knowledge are made by isolated scientists. The aforementioned passage from Galileo's book shows that the philosopher took care to explain his ideas in such a way that as many people as possible could understand them without difficulty. It was not a knowledge for the few.

This same imaginary experiment breaks with the deformed view of Exclusively Analytical Science, according to Cachapuz and collaborators (2005) who point out that this image of science is when it is reduced in a partialised and simplified way without synthesis and a more complex study. Galileo, by considering the conditions of the experiment in the ship's compartment, is making a reductionist assumption that the sea will not shake the ship, not considering air resistance and any dissipative force. However, even disregarding all this, the experiment can be related as if it were on Earth. As Bastos Filho (2005) points out, by making a "falsification" of the real,

When we choose certain elements of reality and, furthermore, disregard some that are not relevant to our study, we are reducing reality. The author gives Galileo-Descartes-Newton's law of inertia as a good example of this reductionism.

At the start of the debate, the students reinforce the idea that it would be much simpler for the Earth to revolve around itself in 24 hours than for the entire universe to revolve around itself in that time. They say this because they are already aware of the concept of speed, distance and time. They didn't need to observe a similar experiment to reach this conclusion. Therefore, this part, even though it is a small excerpt, breaks the deformed view of Empiricist and Atheoretical Science, according to Cachapuz and collaborators (2005) who state that this way of observing Science is when they argue that to learn Science it must be through experiments without needing theoretical knowledge. A neutral observation.

CHAPTER 7

FINAL CONSIDERATIONS

The analysis carried out in the previous section shows that the deformed view of Decontextualised Science is broken down in tables 01 and 04 and their respective transcripts, and in tables 03 and 06. This "break" can be seen when the students/actors realise that Galileo only wrote the book Diálogo because the Pope had given him permission to do so. However, the Pope said that the theories should be explained hypothetically. Galileo did not comply. In addition, Galileo published his book in Florence because if he did it in Rome it would be easier for the Church to interfere in the writing and publication of the work. So there was a relationship between scientific knowledge and society, to the extent that this institution could interfere in the development of science at the time. So much so that it banned the sale of the books and summoned Galileo to appear before the Tribunal of the Inquisition.

Chart 02, 03 and 08 show a break with the deformed view of accumulative science. This can be seen in the students' responses to comments about the characters in the book Galileo's Dialogue who debated the Aristotelian-Ptolemaic and Copernican theories with some of the author's ideas. These theories were in "dispute" at the time as to which one best explained the movement of the stars and physics on Earth. The students also emphasised that Galileo contributed to "breaking" the Aristotelian-Ptolemaic ideas. So there was a dispute between theories.

The deformed view of Aproblematic and Ahistorical Science can be seen in chart 08. This can be understood by looking at the students' comments, which reinforce the idea that scientific knowledge is not absolute truth, and that there is always a new theory that can overturn the previous one. The story of the Theatrical Text turned out to be an example of this, since the students said that just as Galileo's ideas contributed to "breaking" the Aristotelian-Ptolemaic theory, so new ideas can replace his. In addition, there is the assertion that there is other knowledge besides scientific knowledge and that each human being has their own truth that is not necessarily linked to science.

In the last transcript of the previous section, three deformed views of science are broken down. The *Individualist and Elitist view* is "broken" in the students' discussion of Galileo's thought experiment, when they understand the ideas of the situation created by the scientist; showing that science is not a knowledge of the few. The *Exclusively Analytical* is broken by observing Galileo's reductionism in order to understand the experiment in the ship's compartment, which shows the movements of animals and a drop of water in order to understand certain movements on Earth. And the *Empiro-Inductivist and Atheoretical* are "broken" when the students realise that it is easier for the Earth to revolve around itself in 24 hours than it is for the Universe to revolve around itself in 24 hours. They say this because they already have an idea of distance, speed and time. They don't need to carry out the experiment to reach this conclusion.

Based on the aforementioned analyses and considerations of the tables and transcripts, we infer that some deformed views of science, such as the *Decontextualised;* the *Accumulative; the Aproblematic and Ahistorical; the Individualistic and Elitist; the Exclusively Analytical; the Empiricist-Inductivist and Atheoretical*, were debated and, even if partially, these conceptions were broken by most of the students. And, within these discussions, the view that scientific knowledge is absolute truth was also demystified by most of the participants, since they observed throughout history that geocentrism, which was considered the "true" theory of the time, did not explain the discoveries made at the time, and heliocentrism became the most coherent theory with the new discoveries and ideas of the time. Following this example, it can be concluded that working with the History of Science in Science Teaching enables students to realise that their knowledge is not absolute truth, since based on new research, other ideas can be discussed and considered as the "new truth".

From the analyses of the first transcript, tables 05 and 07 and their respective transcripts, and table 06, it can be seen that the Theatrical Text "A Matuta and *The Galileo Affair"* was used by the students/actors to discuss and criticise Galileo's attitudes, such as: publishing the book in Florence instead of Rome, with the intention that his work wouldn't be analysed by people from the Church in the latter city, which is where the Vatican is located; not obeying an order from the Pope, writing the book in a non-hypothetical way; and gaining advantages from the telescope by telling the Venetian government that he invented it, even though he wasn't the creator. Just as they criticised Pope Urban VIII's stance in the *Galileo Affair* for summoning him to the Inquisition for a personal matter, and the Catholic Church for interfering in the development of science, even though it had "proof" in favour of heliocentrism. Matuta's attitude at the end of Scene V, which is explained in question 05, was also debated and criticised by the students. Some students said she was contradictory and others were able to understand her attitude differently. From these observations, it can be concluded that the Text Te atrai "A Matuta e O *Caso Galileu"* can be used as an "action model" for Brecht's Didactic Play.

The activity showed that theatre using Brecht's Didactic Play as a theoretical foundation was an important pedagogical practice for the students'/actors' education in such a way that it provoked reflective debates, leading them to become critical subjects of the attitudes taken by the characters in the play, as well as those of Galileo, Pope Urban VIII and the Catholic Church. In addition, the careful retrieval in the Theatre Text of the important moment in the History of Science known as *The Galileo Affair* was of enormous value in enabling the students/actors to discuss scientific knowledge to the point of becoming capable of breaking down, albeit partially, deformed views of the nature of Science.

Finally, the school, which is a place of learning that provides multiple ways of teaching, makes it possible for Scientific Theatre, in the spirit and orientation of Brecht's Didactic Play, to be an excellent way of highlighting crucial episodes in the history of science. Furthermore, this opportunity also represents an excellent pedagogical practice for the intellectual development of students in order to improve their intellectual formation and understanding, as well as providing the dissemination of scientific knowledge. The discussion, understanding and incentives provided by this activity encourage students to want to learn more and more about the subject.

REFERENCES

ALBUQUERQUE, G. **Dicionário de Termos Nordestinos**. Available at: < http://www.jessierquirino.com.br/2006/dados/dicionario.pdf >. Accessed on: 28 October 2013, 11 p.

BALDOW, R.; MONTEIRO Jr, F. N.. Physics Textbooks and Their Omissions and Distortions in the History of the Development of Thermodynamics. **Alexandria Magazine**, v. 3, n. 1, p. 3-19, 2010.

BALL, D. **Backwards and Forwards: a Guide to Reading Theatre Plays**. São Paulo: Perspectiva, 2ª edition, 2011.

BANFI, A. **Galileo Galilei**. Lisbon: Portugália Editora, 1949.

BARBOSA, A. M. T. B. **Inquietações e Mudanças no Ensino da Arte**. São Paulo: Cortez, 2002.

BARDIN, L. **Content Analysis**. São Paulo: Edition 70, 2011.

BASTOS FILHO, J. B. **Copernicus' Theoretical Reference is Aristotle's Physics**. In: XI Simpósio Nacional de Ensino de Física, Niterói-RJ, p. 420-425, 1995.

BASTOS FILHO, J. B. **Reductionism: an Epistemological Approach**. Maceió: EDUFAL, 2005.

BASTOS FILHO, J. B. Which History and Philosophy of Science Can Improve Physics Teaching? In: Luiz O. Q. Peduzzi; André Ferrer P. Martins; Juliana Mesquita Hidalgo Ferreira (Orgs.). **History and Philosophy of Science in Teaching**. Natal: EDUFRN, p. 65-84, 2012.

BATALHA, R. R. M.; SOUSA Jr, F. S.; SANTOS, A. G. D.; SOUZA L.; FALCONIERI, A. G. F.. **Scientific theatre as a motivational methodology in the science teaching-learning process**. Natal-RN: 4th North-Northeast Chemistry Congress and 2nd North-Northeast Chemistry Teaching Meeting, p. 1-7, 2011.

BIBLE. Portuguese. **Holy Bible**. Translated by Father Antônio Pereira de Figueiredo. São Paulo: Difusão Cultural do Livro, s/a.

BRASIL. **PCN+**: Educational guidelines complementary to the National Curriculum Parameters, Natural Sciences, Mathematics and their Technologies. Brasilia: Ministry of Education and Culture, 2002.

BRECHT, B. **The Life of Galileo**. São Paulo: Editora Abril Cultura, 1977.

BRECHT, B. **Estudos sobre Teatro**. Rio de Janeiro: Nova Fronteira, 1978.

BRECHT, B. **Dialectical Theatre**. Rio de Janeiro: Editora Civilização Brasileira, 1967.

CACHAPUZ, A. et al. **The Necessary Renewal of Science Teaching**. São Paulo: Cortez, 2005.

COHEN, I. B.. **The Birth of a New Physics**. São Paulo: EDART, 1967.

COPÉRNICO, N. **As Revoluções dos Orbes Celestes**. (Translation by A. Dias Gomes and Gabriel Domingues, Introduction and Notes by Luís Albuquerque) Lisbon: Calouste Gulbenkian Foundation, 1984.

COPÉRNICO, N. **Commentariolus:** Nicolaus Copernicus' short commentary on his own hypotheses about celestial movements / Introduction, translation and notes Roberto de Andrade Martins. São Paulo: Editora Livraria da Física, 2ª Edition, 2003.

COSTA, E. B. C.; RIBEIRO, N. B. P.; SOUZA, R. R.. de. **The Use of Theatre to Enrich the Learning of Physics Content in Primary and Secondary Schools**. Rio de Janeiro: XVI National Symposium on Physics Teaching, p. 1-3, 2005.

CRUZ, M. A. S. **Glossary of the Northeastern Sertão**. Available at: < http://sertaodesencantado.blogspot.com.br/2010/04/glossario-sertanejo.html >. Accessed on: 20 Nov. 2013.

DRAKE, S. **Galileo**. Lisbon: Dom Quixote Publications, 1981.

FISCHER, K. **Galileo Galilei**. Barcelona: Editorial Herder, 1986.

FLICK, U. **Introduction to Research Methodology**: A Beginner's Guide. Porto Alegre: Editora Penso, 2013.

GALILEI, G. **Ciência e Fé:** Cartas de Galileu sobre o Acordo do Sistema Copernicano com a Bíblia. (Tradução e Org. Carlos Arthur R. do Nascimento) São Paulo: Editora UNESP, 2ª ed.

GALILEI, G. **Diálogo sobre os Dois Máximos Sistemas do Mundo Ptolomaico & Copernicano**. (Translation, Introduction and Notes by Pablo Rubén Mariconda). São Paulo: Discurso Editorial / Imprensa Oficial do Estado de São Paulo, 2nd edition, 2004.

GALILEI, G. The **Messenger of the Stars**. São Paulo: Duetto Editorial, 2009b.

GARDAIR, T. L. C.; SCHALL, V. T. Ciências Possíveis em Machado de Assis:

Teatro e Ciência na Educação Científica. **Ciência & Educação Magazine**, v. 15, n. 3, p. 695-712, 2009.

GATTI, S. R. T.; NARDI, R.; SILVA, D. History of Science in Physics Teaching: A Study on the Teaching of Gravitational Attraction Developed with Future Teachers. **Revista Investigações em Ensino de Ciências**, v. 15, n. 1, p. 7-59, 2010.

GEYMONAT, L. **Galileo Galilei**. Rio de Janeiro: Nova Fronteira, 1997.

KEARNEY, H. **Orígenes de la Ciencia Moderna**, 1500 - 1700. Madrid: Ediciones Guadarrama, 1970.

KOUDELA, I. D. **Brecht: a Learning Game**. São Paulo: Perspectiva, 2007.

KOUDELA, I. D. **Text and Play: a Brechtian Didactic**. São Paulo: Perspectiva, 2010.

KOUDELA, I. D. **Um Vôo Brechtiano: Teoria e Prática da Peça Didática**. São Paulo: Perspectiva: FAPESP, 1992.

KOYRÉ, A. **The Exact Sciences - The Copernican Revolution**. In: René Taton. General History of Sciences II. São Paulo: Difusão Européia do Livro, v. 1, p. 59-87, 1960.

KOYRÉ, A. **Estudios Galileanos**. Romero de Terreros/Mexico: Siglo Veintiuno Editores, 2009.

MARCENA, A. **Dicionário Escolar da Diversidade Cultural Pernambucana**. Recife: Idea, 2012, 380p.

MACHADO, P. P.; MATOS, W. R. The Use of Theatre in Science Teaching: A Case Study. **Revista Rede de Cuidados em Saúde**, v. 6, n. 1, p. 1-10, 2012.

MARICONDA, P.; LACEY, H. The Eagle and the Starlings: Galileo and the Autonomy of Science. **Revista Tempo Social**, v. 13, n. 1, p. 49-65, 2001.

MARTINS, R. A. What Kind of History of Science Do We Expect in the Next Decades? **Episteme, Filosofia e História das Ciências em Revista**, n. 10, p. 39-56, 2000.

MARTINS, R. A. Galileo and the Rotation of the Earth. **Caderno Catarinense de Ensino de Física**, v. 11, n. 3, p. 196-211, 1994.

MEDEIROS, A. Interview with Kepler: from his birth to the discovery of the first two laws. **Física na Escola**, v. 3, n. 2, 2002.

MEDINA, M; BRAGA, M. O Teatro como Ferramenta de Aprendizagem da Física e de Problematização da Natureza da Ciência. **Caderno Brasileiro de Ensino de Física**, v. 27, n. 2, p. 313-333, 2010.

MINAYO, M. C. S. (Org.). **Social Research: Theory, Method and Creativity**. Petrópolis- RJ: Vozes, 13ª ed., 1999.

MOREIRA, L. M. **Theatre in Museums and Science Centres: a Reading from the Perspective of Scientific Literacy**. 2013. 180f. Thesis (Doctorate) - Faculty of Education, University of São Paulo, São Paulo, 2013.

MOSCHETTI, M. Galileo e as Cartas sobre as Manchas Solares: a Experiência Telescópica contra a Inalterabilidade Celeste. **Especiaria - Cadernos de Ciências Humanas**, v. 9, n. 16, p. 313-340, 2006.

MOURA, D. A.; TEXEIRA, R. R. P. O Teatro Científico e o Ensino de Física - Análise de uma Experiência Didática. **Revista Ciência e Tecnologia**, v. 11, n. 18, p. 1-17, 2008.

MOZENA, E. R. The History and Philosophy of Science in Textbooks on the Problem of Blackbody Radiation (PRCN): Why Not Offer Physics as a

Culture? In: André Ferrer P. Martins (Org.). **Física Ainda é Cultura?** São Paulo: Editora Livraria da Física, 1ª ed., p. 231-258, 2009.

OLIVEIRA, N. R.; ZANETIC, J. **The Presence of Theatre in Physics Teaching**. Jaboticatubas: IX Encontro Nacional de Pesquisa em Ensino de Física, p. 1-12, 2004.

OLIVEIRA, U. A. S. **The Creation of Theatrical Texts from Games and Bertolt Brecht's Didactic Plays**. 2007. 208 f. Doctoral Thesis, Postgraduate Programme in Education, Federal University of Bahia, Salvador, 2007.

OLIVEIRA, U. A. S. M. **Bertolt Brecht's Epic Theatre and Didactic Plays: an Approach to Social Problems and the Search for Political Meaning through Theatre**. Porto Alegre-RS, International Brecht Society Symposium, v. 1, 2013.

ORTEGA Y GASSET, J. **Around Galileo**: A Scheme of Crises. Petrópolis-RJ: Editora Vozes, 1989.

PALMA, C. Arte e ciência no palco (Interview with Luisa Massarani and Carla Almeida). **História, Ciências, Saúde - Manguinhos**, v. 13 (suplemento), p. 233-46, 2006.

PÉREZ, D. G.; Et al. Towards a Non-Deformed Image of Scientific Work. **Revista Ciência e Educação**, v.7, n. 2, p. 125-153, 2001.

PRESTES, M. E. B.; CALDEIRA, A. M. A. **The Importance of the History of Science in Science Education**. Philosophy and History of Biology, v. 4, p. 1-16, 2009.

REDONDI, P. **Galileo Herético**. Madrid: Alianza Editorial, 1990.

RESTON Jr, J. **Galileo - A Life.** Rio de Janeiro: José Olympio Editora, 1995.

SANTILLANA, G. **El Crimen de Galileo - Historia del Proceso Inquisistorial al Genio**. Buenos Aires: Ediciones Antonio Zamora, 1st ed. 1960.

SHEA, W. R. **Galileo's intellectual revolution**. Barcelona: Editora Ariel, 1st ed., 1983.

SILVEIRA, A. F.; ATAÍDE, A. R. P.; FREIRE, M. L. F. Lúdicas Atividades no Ensino de Ciências: uma Adaptação Metodológica através do Teatro para Comunicar a Ciência a Todos. **Revista EDUCAR**, n. 34, p. 251-262, 2009.

SMITH, A. G. R.. **The Scientific Revolution in the 16th and 17th Centuries**. Lisbon: Verbo Lisboa, 1973.

STEINWEG, R. **Indicators of a Path through Baalinesia: Towards an Associal Theatre**. In: Ingrid Dormien Koudela (Org.). Um Vôo Brechtiano: Teoria e Prática da Peça Didática. São Paulo: Perspectiva: FAPESP, p. 47-73, 1992.

TEIXEIRA, F. N. **Didactic Play, Experience and Shock: the** Fatzer **Fragment as a New Way of Narrating**. Porto Alegre-RS: Proceedings of the International Brecht Society Symposium, v. 1, 2013.

VAZ, A. Neither Wise nor Foolish: Passionate about Physics in the Current of Prevailing Cultures. In: André Ferrer P. Martins (Org.). **Física Ainda é Cultura?** São Paulo: Editora Livraria da Física, 1ª ed., p. 175-189, 2009.

VIANNA, D. M. Citizen Education for our Students - a Cultural Context for Physics Teaching. In: André Ferrer P. Martins (Org.). **Física Ainda é Cultura?** São Paulo: Editora Livraria da Física, 1st ed., p. 131-149, 2009.

ZANETIC, J. Physics and Culture. **Revista Ciência e Cultura**, v. 57, n. 3, p. 21-24, 2005.

ZYLBERSTAJN, A. **The Evolution of Conceptions of Force and Movement**. Department of Physics, Federal University of Santa Catarina. Available at: <www.fsc.ufsc.br/~arden/evolucaohist.pdf>. Accessed on: 01 Feb 2011.

APPENDIX A

THE MATUTA AND THE GALILEO CASE

Characters

GALILEO GALILEI

MATUTA, a fictional character who lives with Galileo

CASTELLI, Benedictine priest and friend of Galileo

CARDEAL, a fictional character who is a faithful friend of the Pope

POPE URBAN VIII

MAN OF THE CHURCH, works for the Pope

NICCOLINI, friend of Galileo and Florentine ambassador

INQUISITOR OF FLORENCE

MACULANO, commissary general of the Inquisition

VIRGINIA, daughter of Galileo

CHARLES[13]

ARISTOTLE

TIME: 1632 to 1633; PLACE OF SCENE: Italy

[13] There are ten cardinals taking part in Scene XVII. Two of them also take part in scenes X, XII and XVI.

SCENE I

Galileo in Rome is visited by his friend Casteli.

Galileo: Hello my friend Castelli, how are you?

Castelli: All right, my dear Galileo, illustrious Florentine astronomer and mathematician. We need to talk about the publication of your book.

Galileo: Right! What do you want to talk about?

Castelli: Fr Riccardi has already revised your book, just as Fr Rafael Visconti did the scientific part, and only asked for a few corrections to your work. **Galileo**: Marvellous! We've done a great job.

Castelli: I suggest you travel to Florence to speed up the publication of the book.

Galileo: Why?

Castelli: Because it will be easier for you to publish there. Here in Rome is the Pope and the people here who work for the Catholic Church, and they are very strict. They may not take kindly to the criticism you made of geocentrism in your book and your defence of heliocentrism. You know very well that we're in a time when it's the Church that says what's right or wrong. In Florence the book will be in safer hands.

Galileo: That's true. Here the process for publishing the book would be more rigorous. In Florence I'll have fewer problems.

Castelli: Exactly!

Galileo: Thanks for the tip. I'll go ahead and visit the beautiful city of Florence.

SCENE II

Galileo, already in Florence, is sitting at his desk with a worried look on his face. Matuta arrives and strikes up a conversation:

Matuta: Oxeeeeeente Galileo! Weren't you in Rome getting your book published?
Galileo: Yes, but Castelli suggested that I speed up the process here in Florence. It will be easier and safer to publish the work here.
Matuta: But did you have any trouble in Rome?
Galileo: Not so far. In the revisions that were made there, I was only asked to make a few changes to the book. Not too many problems.
Matuta: It took them ages to authorise the publication of this book. Are you screwed?
Galileo: Well, the work is being revised again. Only by a priest here in Florence. He shouldn't create any problems. But the preface and the final part are being analysed in Rome. That could be the danger.
Matuta: But why should there be any danger? Didn't you approach the two theories hypothetically as the Pope advised? Didn't you put the title as he suggested? Or did you wind it up?
Galileo: With regard to the title, that's fine. I really did follow the Pope's advice. But I created a dialogue between three characters in the book that in the end the arguments used always show that the heliocentric theory is right, and the geocentric one is not.
Matuta: Viiiiiixe Maria, what do you mean?
Galileo: There are three characters. Salviati uses my arguments. He speaks for me; he shows how the heliocentric theory is right. Simplicio, on the other hand, defends the idea of the Peripatetics, those silly ideas. At least at the moment I can say that. He
In the end, he always looks like a fool, his arguments are ridiculed. And the third character, Sagredo, always agrees with Salviati.
Matuta: At the end of the debate it ends up being two against one. And in this case, you didn't do as the Pope suggested. You played him. And that was cheesy.
Galileo: I hope I can open the eyes of these peripatetics. It's not possible! They look

through the telescope and can't see. Or they try to.

Matuta: Oxeeeeeeente, what do you mean?

Galileo: For example, when I showed that the Moon had craters and mountains, Father Clavio, to defend Aristotle's idea of the perfect sky, came out and said that these craters and mountains were filled with a transparent crystalline substance.

Matuta: But that's pushing it. What a stupid idea.

Galileo: Exactly! These Aristotelians keep using passages from the Bible to say that the Earth is at the centre. But who's to say we're not interpreting the Holy Scriptures correctly?

Matuta: It's a possibility.

Galileo: As a good Catholic, I repeat the beautiful words of the wise theologian St Augustine: "Let us neither be seduced by the loquacity of a false philosophy nor be frightened by the superstition of a pretended religion".

Matuta: Now you've messed with my head. Explain these words of St Augustine better.

Galileo: There has to be a balance. If it is being shown that the Copernican theory explains much more of the nature around us and the Aristotelian-Ptolemaic theory has serious flaws and cannot explain facts that have been discovered, then theologians have to make a new interpretation of the Holy Scriptures, so that it is compatible with what is being proven.

Matuta: Got it! But these morons don't want to do that. They think they have the absolute truth. But who are we to determine that? And this Aristotle seems like a God to these people. What a stupid person!

Matuta frowns at Aristotle.

Galileo: Understand Matuta, Aristotle is not the problem. He was a great sage of his time. I admire him deeply. If he were alive today, I'm sure he wouldn't say half of what these peripatetics have been saying.

Matuta: On impulse, we end up talking badly about him. But those Aristotelians in the Church say the right thing.

Galileo: What?

Matuta: They say that heaven is perfect and the imperfections are down here. That's why we see ugly people here. Voooooote!

Galileo: Don't be mean. They're not to blame for being born that way.

Galileo: Returning to my book, if the Church agrees to its publication, I hope to be able to change the world view defended by these peripatetics. I'm very much looking forward to its publication.

SCENE III

Galileo is very agitated and starts shouting for Matuta.

Galileo: Matuta, Matuta, Matuta! Come here!

Matuta: Fuck off! Are you mad?

Galileo: The book has been published and many people are buying it.

Matuta: That means the Church has authorised publication.

Galileo: Exactly!

Matuta: Let me see the book.

Galileo enthusiastically hands the book to Matuta, who is amazed at the size of the work.

Matuta: Come on, what a great book. You must have been hot here with your arguments defending heliocentrism and criticising geocentrism.

Galileo: Just a little. I don't think those peripatetics will continue to defend the existence of the sphere of fixed stars and that it revolves around the Earth in 24 hours.

Matuta: For that to happen, it must be really fast, because that sphere must be a long way away.

Galileo: Exactly. As Copernicus warned us, for these stars to revolve around a fixed Earth in 24 hours, they will tend towards infinite speed.

Matuta: But didn't Aristotle himself not accept the idea of infinity?

Galileo: Following Aristotle's own thinking, Copernicus showed that it was much simpler for the Earth to rotate in 24 hours than for all the stars to circle the Earth in that time.

Matuta: Huuumm! I see! But Galileo, if we threw a stone upwards, shouldn't it fall a certain distance from us, as the Greeks said? It's a great argument against the idea of the Earth's rotation.

Galileo: Matuta, think with me.

Matuta: Right!

Galileo: Imagine you and me inside a compartment on a ship.

Matuta: I'm imagining it.

Galileo: Inside this place there are various flying animals such as flies and butterflies, and a container full of water and small fish inside.

Matuta: So this is where you, me and these animals are?

Galileo: Yes. Also, a bucket full of water stuck up high with a small hole in the bottom letting small drops of water fall into another small bucket below it.

Matuta: I have this scene in my head.

Galileo: Good! But you must first accept that the ship is at rest, and imagine the movement of all these animals and the drops of water.

Matuta: All right!

Galileo: Now let's imagine this ship travelling at a uniform speed.

Matuta: I'm thinking about the scene, but won't the sea water rock the ship?

Galileo: We have to falsify this reality. In our example, the sea doesn't hinder the uniform motion of our ship.

Matuta: Cleaning!

Galileo: In an experiment like this we would see flying animals fly in the same way as in the first example, just as fish swim in the same way. The drops of water will fall the same way into the bucket below.

Matuta: Huuuuum! Very interesting. In this case, the ship is the Earth, we're on top of it watching the movements of these animals and the drop, which continue in the same way.

Galileo: That's the way. The movement of the ship is common to all the things inside it. Just like the Earth and what's on it.

Matuta: You really are a good bitch!

Galileo: I hope that with my book the Aristotelians will be able to understand this kind of idea.

Matuta: I'm delighted that your book has been published. It will teach a lot of people. But I hope your enemies don't turn up bad-mouthing you to the Pope and the cardinals.

Galileo: With the authorisation I've had from the Vatican, there's no need for me to worry.

SCENE IV

A Vatican cardinal goes to meet the Pope to discuss the contents of Galileo's book.

Cardinal: Your Holiness, I've come here to talk about the work that Galileo Galilei has just published entitled Dialogue on the Two Ultimate Ptolemaic and Copernican Systems of the World.

Pope Urban VIII: Right! I even suggested the title.

The Pope raises his head, turns up his nose and is proud of it.

Pope Urban VIII: But is there a problem with this work?

Cardinal: I think everyone.

Pope Urban VIII: What do you mean?

The Pope is astonished by the cardinal's statement.

Cardinal: Galileo openly defended the Copernican system in this book.

Pope Urban VIII: It's not possible! I suggested that he at most approach it hypothetically.

Cardinal: He didn't do it that way. And there's more.

Pope Urban VIII: What else is there?

Cardinal: He created a character in the book called Simplicio who defends Aristotle's ideas. But this character is ridiculed with his arguments. He looks like a fool.

Pope Urban VIII: Still got it!?!?

Cardinal: The Jesuits Scheiner and Grassi are also claiming that Galileo invented this character to ridicule him. That His Holiness is Simplicio himself. So much so that some of the arguments used by this character are the same as those of His Holiness.

Pope Urban VIII: It's not possible that he did something like that. Would he have dared to do that?

The Pope punches the support of the throne on which he is sitting and gets up immediately afterwards.

Pope Urban VIII: I supported him in the 1616 trial, and he comes to do this to me? He's crossed the line.

Cardinal: His Holiness has to see what he's going to do. You know that your current situation is not good. The Spanish cardinals are saying that His Holiness is putting too much pressure on the heretics.

Pope Urban VIII: These Spaniards are like this because I was supported by the philo-French cardinals.

Cardinal: Regardless, at the moment they are putting pressure on His Holiness. Does His Holiness remember what Cardinal Gaspare Borgia did at the Consistory?

Pope Urban VIII: Don't remind me. What an unpleasant situation that was. He attacked me in front of everyone. He said I wasn't tough on heretics.

Cardinal: Exactly, he said that. I'm not trying to put pressure on His Holiness to make a decision. But from what I understand, His Holiness has been mocked, called a fool, silly, that his arguments are weak, he's gone over his head,...

The Pope interrupts, shouting.

Pope Urban VIII: That's good! That's good! Got it!

The Cardinal is startled to the point of falling over.

Pope Urban VIII: I will arrange for a commission to analyse this book, and I will ask them to collect all the works that are on sale. Thank you very much for this information. God bless you!
Cardinal: Amen!

SCENE V

Galileo is sitting down, while Matuta is fiddling with the telescope. Noticing Galileo's worried face, Matuta goes towards him and starts a conversation:

Matuta: Why do you look so upset?
Galileo: Dear Matuta, I've just read a letter from Father Campanella confirming the bad news.

Galileo shakes his head in disapproval.

Matuta: Apart from the books that were collected here by the inquisitor of Florence, is there anything else?

Galileo: The books were collected everywhere. And the Pope set up a commission to analyse the work.
Matuta: Viiiiiiiiixe! Will you have to go to Rome?
Galileo: It's likely. That's all you need, at that age having to go through an Inquisition.

Matuta: But are you going to turn into toast?

Galileo is astonished by what Matuta says and replies:

Galileo: Matuta, for God's sake, don't even say that.
Matuta: It's because I remembered Giordano Bruno.
Galileo: Forget about Bruno. I have to talk to the authorities and important people I trust to help me. It doesn't look like it's going to be easy to get out of this situation. What will this commission say about my book?
Matuta: Why do these Church people have the authority to say what is right and wrong? Why have they acquired so much power that they can forbid a person to publish their ideas? Why can they torture and burn people alive? I consider all of this to be absurd! Do you know what I'm going to do?
Galileo: What?

Matuta looks at Galileo's face, makes a bit of a mystery and says:

Matuta: I'll pray for you.
Galileo: Pray? Pray? Yes! Maybe God will enlighten me.

SCENE VI

The cardinal talks to the Pope again about the analysis made of Galileo's book "Dialogue" by the commission set up by the Pontiff.

Cardinal: Your Holiness, the commission that was set up to analyse Galileo's book had five sessions analysing the work and found some problems with it.
Pope Urban VIII: Well done! Tell us the main ones she found.
Cardinal: Firstly, Galileo didn't treat the Copernican theory as hypothetical, but clearly defended the mobility of the Earth and the stability of the Sun.
Pope Urban VIII: That's good to know. And what else?

Cardinal: Secondly, he described the phenomenon of the tides by relating it to the stability of the Sun and the movement of the Earth; this explanation is erroneous in the eyes of the science defended by the Church.

With a wry smile, the Pope comments:

Pope Urban VIII: His situation is not good.

At this point, the Pope shows that he is determined to take revenge on Galileo. This information makes him happy, as it sets the precedent for making Galileo's situation worse.

Pope Urban VIII: Anything else to make his case worse?
Cardinal: Galileo bypassed the order issued by the commission of the Holy Office in 1616.
Pope Urban VIII: You don't need to tell me any more. This last one alone is reason enough to summon you to the Inquisition.

Cardinal: But the commission that analysed the book said that with a few corrections we wouldn't have a problem with this work.

The Pope makes it clear that he doesn't want to change the book, he wants to punish Galileo, he wants revenge.

Pope Urban VIII: The commission has played its part brilliantly. But let me sort it out now. I'm going to speak to Cardinal Antonio Barberini to give the inquisitor of Florence the task of informing Galileo that he has to be transferred to Rome quickly. Let's see how he copes.

SCENE VII

Niccolini arrives at his scheduled audience with the Pope.

Man of the Church: His Holiness, Niccolini has arrived.
Pope Urban VIII: Send him in.

Niccolini enters and starts the conversation.

Niccolini: Your Holiness, I would like to thank you for the space.
Pope Urban VIII: What do you want?

Niccolini realises that the Pope isn't very happy with Galileo and that the conversation isn't going to be easy.

Niccolini: I'm here to deliver a note of protest from the government of Florence against the commission that has been set up to analyse Galileo's book...

The Pope interrupts, speaking loudly.

Pope Urban VIII: How dare you come here to defend this Galileo? How dare you bring this note here? Galileo did not fulfil an order from the Holy Office.
Niccolini: Your Holiness, the Church itself authorised the publication of the book...

The Pope interrupts again.

Pope Urban VIII: Galileo deceived us! He dared to enter into subjects he shouldn't have. Into serious and dangerous content. He approached a heretical theory recklessly. Don't come to Galileo's defence.
Niccolini: But Fr Riccardi reviewed the work and saw no faults. He authorised the

book to be published.

Pope Urban VIII: This is another one who didn't do his job properly. And he even authorised the printing of this book in Florence.

Niccolini: Your Holiness, shouldn't Galileo be informed of the problems that are being said about his work?

Pope Urban VIII: We don't usually work like that. We don't anticipate anything. And Galileo is well aware of the problems he has in his book.

Niccolini realises that the Pope believes Galileo did everything in bad faith.

Niccolini: Your Holiness...

Once again the Pope interrupts Niccolini.

Pope Urban VIII: Subject closed. We have nothing more to say about it. Galileo will be informed of the decisions the Church is taking.

Niccolini: Thanks for the space.

Pope Urban VIII: God bless you!

Niccolini: Amen!

Niccolini leaves the room.

SCENE VIII

In Galileo's house...

Matuta: Galileo, there's a goat who wants to talk to you.

Galileo: Who is it?

Matuta: He said he was the inquisitor of Florence.

Galileo: Send him in.

Matuta goes out to call the inquisitor of Florence. He enters and goes to talk to Galileo.

Inquisitor of Florence: Hello Galileo!

Galileo: Hello dear friend, any news from Rome?

Inquisitor of Florence: Yes, I have. I've just received a letter from Barberini. It seems that the situation is not very good.

Galileo: And what does he say in the letter?

Inquisitor of Florence: He said you should go to Rome soon.

Galileo: But I'm not well; I'm ill. Please ask me to postpone this trip. I need a few more days.

Inquisitor of Florence: I can no longer make this request.

Galileo: Why?

Inquisitor of Florence: They're already saying that you're abusing their goodwill. And they won't tolerate any more delays on this journey. They have threatened, if you don't go now, to send a commissioner with a doctor to come and pick you up and take you straight to the prison of the supreme court.

Galileo: I have no choice. I'll have to go to Rome.

Inquisitor of Florence: It's the most sensible decision at the moment.

Galileo: Thanks for the information. I hope my friend Niccolini has some good news for me.

Inquisitor of Florence: Good luck in Rome.

Galileo: Thank you!

Inquisitor leaves.

SCENE IX

Galileo receives a visit from Niccolini.

Galileo: My friend Niccolini, what a pleasure to see you again.
Niccolini: You know it's always an honour to talk to you, but I don't have any good news.
Galileo: Have you spoken to the Pope?

Niccolini: Yes, he's very upset about the content of your book, mainly because you talked about the Copernican theory in a non-hypothetical way. Thus complicating the science defended by the Church.
Galileo: I can't stand these peripatetic views any longer. They insist on being blind. They observe the sky through a telescope, but they continue with their blindness. Tell me: how does the Ptolemaic theory explain the phases of Venus?
Niccolini: I know there's no way, but we can't think about that now, but about the inquisition you'll have to face.
Galileo: I can't believe the Pope has gone that far. And I even thought that with Urban VIII in power at the Vatican, everything would change for the better.
Niccolini: They're turning his mind against you. Grassi and Scheiner, for example, are trying to get it into his head that the character Simplicio is the Pope himself. And that you wanted to ridicule Urbano through this character.
Galileo: That's absurd! How can the Pope believe that? I have to talk to some people close to him to get it out of his head. It's ridiculous.
Niccolini: I think so too. You could talk to the Venetian government to help you. Didn't you show them the telescope you invented?
Galileo: Invented by whom?
Niccolini: For you, wasn't it?

Galileo laughs and says:

Galileo: Not for me.

Niccolini: But you told them that.

Galileo: Yes.

Niccolini: And why did you say that?

Galileo: What I wanted was money! I couldn't do any research with all the private lessons I had to give to pay my bills. I had to pay for my two sisters' weddings, I helped my mum and brother who loved to spend and I had to support my three children.

Niccolini: Galileo, you surprise me every day.

Galileo: Rubbish. It's nice to have that kind of winding up.

Galileo: Now that Scheiner won't leave me alone. He's been annoyed with me ever since I said I'd observed the sunspots. He kept saying that he made this observation first, and that I was informed of it. Now he's invented this lie to the Pope.

Niccolini: But who saw the sunspots first?

Galileo: You think I'm going to say he did it? Never mind that.

Niccolini: But did you see the four stars[14] that revolve around Jupiter first?

Galileo: Of course. No doubt about it. But let's get back to our subject. **Niccolini**: Well, you'll have to go to the Holy Office.

Galileo: Right! I don't have a choice.

SCENE X

Galileo arrives at his first meeting with the Inquisition and enters a room with three people sitting at a table in front of them. Two cardinals at the ends and Maculano in the middle. This is the man in the Church responsible for coordinating this process. He asks Galileo to sit down, and then the interrogation begins.

Maculano: Come in. Sit in this Galileo Galilei chair.

Galileo: Thank you very much!

---

[14] These stars are the satellites of Jupiter that Galileo observed with his telescope.

Galileo sits down in a chair right in front of Maculano.

Maculano: Do you know why you're here?
Galileo: I think it's because of the book I recently wrote.

Maculano picks up the book on the table and continues the conversation:

Maculano: Here's the book. Can you tell us if it's yours?

Galileo picks up the book, opens it, analyses it and replies.

Galileo: This is my book. It was published last year.
Maculano: Did you know that he defends theories that have been banned from being taught since 1616?
Galileo: Can you clarify what you mean?
Maculano: Before that, tell me what you were told, and what document did you receive from the Church in 1616?
Galileo: I received a document saying that I could no longer support or defend the immobility of the Sun and the movement of the Earth. But there was nothing to stop me talking about these ideas hypothetically.
Maculano: I have a document here that clearly states that the immobility of the Sun and the movement of the Earth should not be supported, defended or taught in any way.
Galileo: I wasn't told that these ideas couldn't be taught in any way.

Maculano: But didn't Cardinal Bellarmine inform you personally?
Galileo: When Cardinal Bellarmine spoke to me, he said that the immobility of the Sun and the movement of the Earth were ideas contrary to the Holy Scriptures, and that they were forbidden to be defended and supported.
Maculano: And in your book, did you support or defend Copernican ideas? **Galileo**: At no point in my book did I defend or support the ideas of Copernicus. On the

contrary, they were shown to be invalid and inconclusive.

Maculano shows a little anger at hearing these words, since Galileo was blatantly lying.

Maculano: In that case, I'm going to ask for a commission to re-evaluate your book to make sure you're right. Go ahead.

As soon as Galileo leaves, Maculano speaks up:

Maculano: He'll say what we want. I'll convince him.

SCENE XI

Maculano goes to meet Galileo to try to convince him to confess his mistakes.

Maculano: Hello Galileo, how are you?
Galileo: Hello Maculano, what a surprise to see you here. What can I do for you?
Maculano: I've come here to talk about the accusations being levelled at you.

Galileo: Right! What do you want to talk about?
Maculano: Although you said in your interrogation that you didn't support or defend the Copernican theory in your book, we know that's not true. Even the theologians who analysed your book made it very clear that you defended the immobility of the Sun and the movement of the Earth.
Galileo: With all due respect to these theologians, but I don't agree with the conclusion they've reached.
Maculano: Dear Galileo, I'm not going to insist on this. But let me make it clear: no matter what you argue, you won't be able to convince the Pope or me. You will only delay this process further. And that could be bad for you.

Galileo: Many of my enemies have turned the Pope's head. That's why he's acting this way.

Maculano: If I were you, I wouldn't insist. It would be better for you to confess your mistakes in that book straight away.

Galileo: But there are no mistakes in the book. I only discussed the two systems hypothetically.

Maculano: You're not going to convince me. As I've already told you, we know that you have defended ideas that are contrary to the Holy Scriptures. You'd better confess your mistakes yourself. I wouldn't want to use rigid methods to solve your case.

Galileo: Is that a threat?

Maculano: It's just a warning about what can happen if you don't want to contribute to the Church. I came here to help you. So that the worst doesn't happen to you.

Galileo: I need some time to think. In the next interrogation, I'll contribute to you.

Maculano: I hope you've understood. May God bless you.

Galileo: Amen!

SCENE XII

Galileo arrives for his second meeting with the Inquisition. Maculano asks Galileo to sit down, and then the interrogation begins.

Maculano: Dear Galileo, come in and sit down.

Galileo: Thank you!

Galileo sits down. Maculano, intent on getting Galileo to confess straight away, asks him a simple, straightforward question.

Maculano: Do you have anything to say?

Galileo: Yes.

Maculano: Feel free to talk.

Galileo: I hadn't read my book for three years. And now that I've reread it, I feel like

I'm reading someone else's work.

Maculano: Yes, but she's yours.

Galileo: Exactly. Although I wrote this book without pretending to defend the Copernican theory, in some passages of my work, a more ignorant reader could be led to believe in ideas related to heliocentrism.

Maculano: Is that a confession?

Galileo: Yes.

Maculano: Be clearer.

Galileo: My mistake...

As Galileo breathed, Maculano was already hopeful that the inquisition would end with that confession.

Galileo: ...was to boast with my words. I ended up boasting about the way I wrote this book.

Maculano was clearly disappointed by this confession. It wasn't what he had expected, nor what he had suggested to Galileo. He still tries to get something out of the scientist.

Maculano: Is that the confession you have to tell us?

Galileo: Yes.

Maculano: Don't you have anything else to say about your book?

Galileo: No. That's the mistake I realised.

Maculano angrily tells Galileo to leave the room.

Maculano: You're free to go.

Galileo: Thank you very much!

SCENE XIII

Maculano goes to meet the Pope.

Maculano: Your Holiness, I've come here to talk about the case of Galileo Galilei.
Pope Urban VIII: Was he able to solve this problem? Did he confess what he defended in the book?

Maculano: That's what I'd like to talk about. In this last interrogation, Galileo made a confession. But it wasn't the one we expected.
Pope Urban VIII: What do you mean? Did he confess or not?
Maculano: He said that after re-reading the book he realised that he had been boasting. His confession was that he boasted about the way he wrote this work.
Pope Urban VIII: This Galileo seems to be toying with us. He doesn't realise how bad his situation is.
Maculano: What does His Holiness propose to do?
Pope Urban VIII: Make him confess that he defended the movement of the Earth and the immobility of the Sun.
Maculano: But if he doesn't?
Pope Urban VIII: He'll have to do it. In any case. Even if we have to use more rigorous methods.
Maculano: Got the message.
Pope Urban VIII: I see that Niccolini is going to come to me to talk about this case.
Maculano: They're realising that it's getting difficult for them.
Pope Urban VIII: Do what I've asked you to do, and we'll be done with this affair in no time.
Maculano: I'm coming, Your Holiness.
Pope Urban VIII: Stay with God.
Maculano: Amen!
SCENE XIV

Before his final interrogation, Niccolini was granted an audience with Pope Urban VIII.
Niccolini: Your Holiness, thank you very much for taking the time to talk to us.

Pope Urban VIII: All right! Did you ask for this audience to talk about Galileo?
Niccolini: Yes.
Pope Urban VIII: Right! Go ahead.

With these words, the Pope showed Niccolini that he didn't want to waste too much time talking about Galileo.

Niccolini: Your Holiness, I come here to ask for your clemency with regard to Galileo.

The Pope makes it clear straight away what is going to happen to Galileo without wanting to talk too much.

Pope Urban VIII: Dear Niccolini, Galileo has ignored orders that he received in 1616. That alone is reason enough for him to be imprisoned.
Niccolini: I ask you to reconsider, Your Holiness. Galileo was always a good Catholic. He was always at the service of the Church.
Pope Urban VIII: Niccolini, Galileo defended in his book the immobility of the Sun and the movement of the Earth. He also ridiculed Ptolemaic ideas, which are defended by the Church, by me, and which are in accordance with Sacred Scripture. For this reason, Galileo will be forbidden to give his opinion on Copernicus' ideas. He will be informed of these decisions at the next interrogation.

Niccolini realises that the decisions have already been made. And that the Pope isn't going to change them.

Niccolini: Right! Thank you for the space. But I hope that by the next interrogation, His Holiness will be able to change this decision. Thank you very much!
Pope Urban VIII: God bless you!
Niccolini: Amen!

SCENE XV

Niccolini goes to meet Galileo.

Galileo: Hello Niccolini, any news?
Niccolini: I spoke to the Pope.
Galileo: And what did he say? After this last interrogation, will they end this process?
Niccolini: No. There will still be one more interrogation.
Galileo: What more do they want?
Niccolini: They'll want a confession from you. Think about what you're going to say at your next meeting.
Galileo: Niccolini, I'm old, I can't take it anymore.
Niccolini: I know, Galileo, but there's not much we can do. The Pope says you'll be banned from talking about Copernican ideas.
Galileo: I expected that decision.

Even though he was aware of the decision to imprison him, Niccolini prefers not to talk.

Niccolini: I hope that's their only decision. Think about what you're going to say tomorrow. You're trapped.
Galileo: That's true. But did he ever say if I'm going to become toast?

Nicollini is astonished by Galileo's question.

Niccolini: Forget that idea. Relax a bit and think about what you're going to say.
Galileo: I'll think about the best way to get rid of this problem. Thank you very much!
Niccolini: You're welcome. I hope everything goes well during this next interrogation.

SCENE XVI

Galileo arrives for his last interrogation by the Inquisition. Maculano, without much procession, asks Galileo to sit down.

Maculano: Come in and sit down.
Galileo: Thank you!

Galileo realises from Maculano's face and the way he welcomes him that he hasn't come to talk much.

Maculano: This is already our third meeting and you still haven't told us clearly what you think about the immobility of the Sun and the movement of the Earth. So I'm going to be very direct and I hope you'll give me a convincing answer. Not least because I hope I won't have to use different methods in this inquisition.

Galileo sighs heavily. He realises that the Church is going to try to get what it wants out of him; it could even lead to torture.

Galileo: I'm here to help.
Maculano: Do you or have you ever supported the immobility of the Sun and the movement of the Earth?

Galileo realises that Maculano has got straight to the point, showing that he doesn't have much patience.

Galileo: In the past, I thought that both Ptolemy's and Copernicus' theories were worthy of discussion. But the wisdom of the authorities has opened my eyes to the fact that Ptolemy's ideas are the only ones that are true and worthy of being taught and discussed.

Maculano: And with regard to your book, at any point did you defend Copernicus' ideas?

Galileo realises he's being backed into a corner. The situation is becoming quite critical.

Galileo: In my book, I didn't support Copernicus' ideas. What I tried to do in that book was to show the arguments that exist in favour of the Ptolemaic theory, as well as the arguments in favour of the Copernican theory. The aim is to show that neither argument has the demonstrative power to reach a conclusion. Consequently, we have to resort to the decision of a higher teaching. That's why I support the decisions made by the Church.

From beyond, Aristotle appears in the sky shouting at Galileo:

Aristotle: Galileo! The Earth is spinning! The Earth is spinning!

Everyone is startled by the shouting.

Maculano: What is it? Is there anyone in this room?
Galileo: No. I don't see anyone.

Galileo looks up and sees Aristotle. Faced with the situation, Galileo speaks very quietly to Aristotle:

Galileo: Shut up, Aristotle! Go away! Don't be in the way any more.

Maculano refocuses on the interrogation and even though he is not completely satisfied with Galileo's statement, he realises that the case can be closed.

Maculano: Anything else to add?

Galileo: No.

Maculano: You'll be called to make the abjuration. You may leave.

SCENE XVII

Galileo enters a room containing some cardinals. He walks towards the Bible on the table, as well as the paper containing his abjuration. Until he reaches the table, everyone is startled by a noise that everyone hears without knowing where it's coming from. Some tiles start to fall to the floor, making a frightening noise. Only Galileo isn't frightened by anything. He continues on his way to the Bible. When he picks it up, so does the paper with his abjuration,

the noises suddenly stop. He kneels with great difficulty in front of the cardinals and begins to read his abjuration:

Galileo: I, Galileo, son of Vincentio Galilei, of Florence, at the age of 70, personally constituted in judgement and kneeling before you, Most Eminent and Most Reverend Cardinals, in the whole Christian republic general inquisitors against heretical depravity, having before my eyes the sacrosanct Gospels, which I touch with my own hands, swear that I have always believed, believe now, and with God's help will believe in the future, all that the Holy Catholic and Apostolic Church affirms, preaches and teaches. Because I, after having been legally summoned that I must totally abandon the false opinion that the Sun is the centre of the world and does not move, and that I could not support, defend or teach the said doctrine in any way, either orally or in writing, and after having been notified that the said doctrine is contrary to Sacred Scripture, written and printed a book in which I treat of the same doctrine already condemned and adduce reasons very effectively in its favour, without presenting any solution, I have been judged vehemently suspected of heresy, that is, of having affirmed and believed that the Sun is the centre of the universe and immovable and the Earth is not the centre and moves. Therefore, wishing to remove from the mind of Your Eminence and all the Christian faithful this vehement suspicion, rightly conceived, with a sincere heart and unfeigned faith, I abjure, curse and detest the aforementioned errors and heresies and, in general, any and all errors, heresies and sects contrary to the Holy Church; and I swear that in future I will never again affirm, either orally or in writing, such things as might subject me to similar suspicions. So help me God and these holy Gospels of yours, which I touch with my own hands .[15]

When he has finished reading, Galileo's eyes show disapproval and sadness at the moment, but he signs the abjuration and then tries to get up, having difficulty doing so because of his poor health. Three cardinals help him up. He places the abjuration paper on the table and before leaving the room, he speaks quietly to himself:

Galileo: What will my life be like now?

SCENE XVIII

Galileo is standing in his room when his daughter Virginia enters.

Virginia: Hello Dad, sit down here for a while and rest.

Galileo, who is almost blind, sits down in his chair with the help of his daughter Virginia, who starts talking to him.

Virginia: Are you all right?

Galileo: I can't get well, my child.

Virginia: But you have to forget that. Seek a more spiritual life. Forget those studies. They'll do you harm.

Galileo: Studies have never hurt me. What hurts me is seeing the truth and going blind.

Virginia: I know that science is everything to you. But you only live sick, you're almost blind, you're almost seventy. Try to think about yourself, about your health.

---

[15] This Abjuration is not complete, as can be found on pages 232 and 233 of Banfi (1949). Some cuts have been made.

Galileo: Science is my health. I have so much to say about physics and astronomy and they're silencing me. That's what makes me sick. But I have to find a way to keep studying.
Virginia: But Dad, you won't be able to publicise these ideas.
Galileo: Nothing I can do will be done quickly. Then I'll have time to think about how I'm going to publish my ideas. I hope I'll have time for that too.

Matuta, who was visiting, hears Galileo's last words and excitedly enters the room.

Matuta: So you mean you denied heliocentrism in order to study new ideas? It was all your doing?
Galileo: Play? What do you mean, played?
Matuta: And I thought you were a slacker. But you didn't stand up to the Church in order to spread new ideas.
Galileo: I didn't confront the Church because I was afraid.
Matuta: Afraid? But they didn't even torture him.
Galileo: They didn't have to. All they had to do was threaten and I did what they wanted.
Matuta: You're too soft.[16]

Matuta leaves the room angrily.

Virginia: Never mind, Dad. You're not going to talk about heliocentrism, are you?
Galileo: I probably won't study it. Then I'll find a dragon on the moon and it'll be a big deal.

Virginia: It would be hilarious, to say the least. Dad, I hope you think carefully about what you're going to do. I'll always be by your side.

---

[16] This conversation between Galileo and Matuta was based on a dialogue between the characters Galileo and Andrea from Brecht's book (1977).

Galileo: Thank you very much, my child. I know I wasn't much of a father to you. But you're being a great daughter to me.

Virginia: I've always loved you, Dad. I'll pray to you.

Virginia leaves the room. Soon afterwards, Galileo gets up and starts talking:

Galileo:

In the game with the Church
I didn't know how to play
She won and I was arrested
I went blind and couldn't see

Not just with the senses
But also with reason
The Holy Scriptures
They didn't accept the sun I saw
Only Gibeon

They made my sun square
Not being able to express myself
If Eratosthenes were alive
Even he couldn't bend

The Earth is still at the centre
Even though the Church and I know that

But the faith that moves mountains
He went beyond reason

I wish I'd had the strength,

Courage and malice
But when they threatened with thorns
Aristotle and Ptolemy became my divas

I'm here today
Not with the same fate as Bruno
But with the truth withheld
Arrested and humiliated
But knowing that one day everything I've done
It won't be death, but life

END

Kolibri 360 GmbH
Ferdinand-Jühlke-Straße 7
99095 Erfurt, Deutschland
produktsicherheit@kolibri360.de

Druck:
CPI Druckdienstleistungen GmbH
im Auftrag der
Kolibri 360 GmbH
Ferdinand-Jühlke-Str. 7
99095 Erfurt